# An Introduction to Geology

# An Introduction to Geology

Eden Wilkins

Larsen & Keller
www.larsen-keller.com

An Introduction to Geology
Eden Wilkins
ISBN: 978-1-64172-085-4 (Hardback)

© 2019 Larsen & Keller

# ▤ Larsen & Keller

Published by Larsen and Keller Education,
5 Penn Plaza,
19th Floor,
New York, NY 10001, USA

**Cataloging-in-Publication Data**

An introduction to geology / Eden Wilkins.
    p. cm.
Includes bibliographical references and index.
ISBN 978-1-64172-085-4
1. Geology. 2. Earth sciences. I. Wilkins, Eden.
QE26.3 .I58 2019
551--dc23

For more information regarding Larsen and Keller Education and its products, please visit the publisher's website www.larsen-keller.com

# Table of Contents

# Preface

Geology is the scientific study of the Earth's surface, its evolution and the processes that have led to its change. The demonstration of the age of the Earth, chronicling of the Earth's geological history, evidence for plate tectonics, and the understanding of past climates have been possible because of advancements in the field of geology. Rock analysis is the most significant area of geological studies. Rock can be of three types namely, sedimentary, igneous and metamorphic. The techniques used in geological investigations are fieldwork, chemical analysis, numerical modeling, rock description and physical experimentation. Hydrocarbon and mineral exploration, hydrological studies, understanding of natural hazards and past climates, etc. are explored from within the framework of geology. This textbook is a valuable compilation of topics, ranging from the fundamental to the most complex theories and principles in the field of geology. It further elucidates the techniques and applications of geology in a multidisciplinary manner. The book strives to be a complete source of information for all students who are looking for an elaborate reference text on geology.

A foreword of all Chapters of the book is provided below:

**Chapter 1**- Geology is a discipline of earth science which studies and analyzes the surface of the Earth, especially the formation of rocks, their composition, and their change overtime. This is an introductory chapter that elucidates the varied processes and mechanisms associated with this area of study, besides delving into the essentials of the history of geology and geologic time scale; **Chapter 2**- As a sub-discipline of geology, petrology refers to the study of the formation of rocks. Petrology is a vast subject that branches out into significant sub-disciplines, such as igneous petrology, sedimentary petrology and metamorphic petrology, which have been thoroughly discussed in this chapter. It also covers in extensive details the fundamental rock types and the concept of rock cycle; **Chapter 3**- Structural geology is the study of the measurements of rock geometries in order to gain insights into their deformation histories, and understand the stress fields. Various concepts and methods, which have been thoroughly explained in this chapter, include outcrop, layering, intrusion, deformation, strike and dip, and rock structures, to provide an extensive understanding of the subject; **Chapter 4**- Any chemical compound, which is naturally occurring and crystalline in nature, is called a mineral. Unlike rocks, minerals have specific chemical compositions and their study is referred to as mineralogy. This chapter discusses the various properties of minerals like their color, streak, hardness, crystal form, etc. as well as the geological processes of their formation.

I would like to thank the entire editorial team who made sincere efforts for this book and my family who supported me in my efforts of working on this book. I take this opportunity to thank all those who have been a guiding force throughout my life.

Eden Wilkins

# Understanding Geology

Geology is a discipline of earth science which studies and analyzes the surface of the Earth, especially the formation of rocks, their composition, and their change overtime. This is an introductory chapter that elucidates the varied processes and mechanisms associated with this area of study, besides delving into the essentials of the history of geology and geologic time scale.

Geology, the fields of study concerned with the solid Earth. Included are sciences such as mineralogy, geodesy, and stratigraphy.

A rock formation called "faux amphibolite," from the Nuvvuagittuq greenstone belt in northern Quebec, dated to 4.28 billion years ago.

An introduction to the geochemical and geophysical sciences logically begins with mineralogy, because Earth's rocks are composed of minerals—inorganic elements or compounds that have a fixed chemical composition and that are made up of regularly aligned rows of atoms. Today one of the principal concerns of mineralogy is the chemical analysis of the some 3,000 known minerals that are the chief constituents of the three different rock types: sedimentary (formed by diagenesis of sediments deposited by surface processes); igneous (crystallized from magmas either at depth or at the surface as lavas); and metamorphic (formed by a recrystallization process at temperatures and pressures in the Earth's crust high enough to destabilize the parent sedimentary or igneous material). Geochemistry is the study of the composition of these different types of rocks.

Pahoehoe lava flow, Kilauea volcano, Hawaii

During mountain building, rocks became highly deformed, and the primary objective of structural geology is to elucidate the mechanism of formation of the many types of structures (e.g., folds and faults) that arise from such deformation. The allied field of geophysics has several subdisciplines, which make use of different instrumental techniques. Seismology, for example, involves the exploration of the Earth's deep structure through the detailed analysis of recordings of elastic waves generated by earthquakes and man-made explosions. Earthquake seismology has largely been responsible for defining the location of major plate boundaries and of the dip of subduction zones down to depths of about 700 kilometres at those boundaries. In other subdisciplines of geophysics, gravimetric techniques are used to determine the shape and size of underground structures; electrical methods help to locate a variety of mineral deposits that tend to be good conductors of electricity; and paleomagnetism has played the principal role in tracking the drift of continents.

Geomorphology is concerned with the surface processes that create the landscapes of the world—namely, weathering and erosion. Weathering is the alteration and breakdown of rocks at the Earth's surface caused by local atmospheric conditions, while erosion is the process by which the weathering products are removed by water, ice, and wind. The combination of weathering and erosion leads to the wearing down or denudation of mountains and continents, with the erosion products being deposited in rivers, internal drainage basins, and the oceans. Erosion is thus the complement of deposition. The unconsolidated accumulated sediments are transformed by the process of diagenesis and lithification into sedimentary rocks, thereby completing a full cycle of the transfer of matter from an old continent to a young ocean and ultimately to the formation of new sedimentary rocks. Knowledge of the processes of interaction of the atmosphere and the hydrosphere with the surface rocks and soils of the Earth's crust is important for an understanding not only of the development of landscapes but also (and perhaps more importantly) of the ways in which sediments are created. This in turn helps in interpreting the mode of formation and the depositional environment of sedimentary rocks. Thus the discipline of geomorphology is fundamental to the uniformitarian approach to the Earth sciences according to which the present is the key to the past.

Geologic history provides a conceptual framework and overview of the evolution of the Earth. An early development of the subject was stratigraphy, the study of order and sequence in bedded sedimentary rocks. Stratigraphers still use the two main principles established by English engineer and surveyor William Smith, regarded as the father of stratigraphy: (1) that younger beds rest upon older ones and (2) different sedimentary beds contain different and distinctive fossils, enabling beds with similar fossils to be correlated over large distances. Today biostratigraphy uses fossils to characterize successive intervals of geologic time, but as relatively precise time markers only to the beginning of the Cambrian Period, about 540,000,000 years ago. The geologic time scale, back to the oldest rocks, some 4,280,000,000 years ago, can be quantified by isotopic dating techniques. This is the science of geochronology, which in recent years has revolutionized scientific perception of Earth history and which relies heavily on the measured parent-to-daughter ratio of radiogenic isotopes.

Paleontology is the study of fossils and is concerned not only with their description and classification but also with an analysis of the evolution of the organisms involved. Simple fossil forms can be found in early Precambrian rocks as old as 3,500,000,000 years, and it is widely considered that life on Earth must have begun before the appearance of the oldest rocks. Paleontological research of the fossil record since the Cambrian Period has contributed much to the theory of evolution of life on Earth.

Several disciplines of the geologic sciences have practical benefits for society. The geologist is responsible for the discovery of minerals (such as lead, chromium, nickel, and tin), oil, gas, and coal, which are the main economic resources of the Earth; for the application of knowledge of subsurface structures and geologic conditions to the building industry; and for the prevention of natural hazards or at least providing early warning of their occurrence.

Astrogeology is important in that it contributes to understanding the development of the Earth within the solar system. The U.S. Apollo program of manned missions to the Moon, for example, provided scientists with firsthand information on lunar geology, including observations on such features as meteorite craters that are relatively rare on Earth. Unmanned space probes have yielded significant data on the surface features of many of the planets and their satellites. Since the 1970s even such distant planetary systems as those of Jupiter, Saturn, and Uranus have been explored by probes.

Some of the important events in the Earth's history are floods, volcanic eruptions, earthquakes, orogeny (mountain building), and plate tectonics (movement of continents).

Geology is divided into special subjects that study one part of geology. Some of these subjects are:

- Geomorphology – the study of the shape (morphology) of the surface of the Earth.

- Historical geology – the history of the events that shaped the Earth over the last 4.5 million years.

- Hydrogeology – the study of water under the surface of the Earth.

- Palaeontology – the study of fossils.

- Petrology – the study of rocks how they form and where they are from.

- Mineralogy – the study of minerals.

- Sedimentology – the study of sediments.

- Stratigraphy – the study of layered sedimentary rocks and how they were deposited.

- Structural geology– the study of folds and faults and how mountains are formed by uplift.

- Volcanology – the study of volcanoes on land or under the ocean.

- Seismology – the study of earthquakes and strong ground-motion.

- Engineering geology -the study of geologic hazards (such as landslides and earthquakes) applied to civil engineering.

- Petroleum geology- the study of petroleum deposits in sedimentary rocks.

## Types of Rock

Rocks can be very different from each other. Some are very hard and some are soft. Some rocks are very common, while others are rare. However, all the different rocks belong to three categories or types, igneous, sedimentary and metamorphic.

- Igneous rock is rock that has been made by volcanic action. Igneous rock is made when the lava (melted rock on the surface of the Earth) or magma (melted rock below the surface of the Earth) cools and becomes hard.

- Sedimentary rock is rock that has been made from sediment. Sediment is solid pieces of stuff that are moved by wind, water, or glaciers, and dropped somewhere. Sediment can be made from clay, sand, gravel and the bodies and shells of animals. The sediment gets dropped in a layer, usually in water at the bottom of a river or sea. As the sediment piles up, the lowers layers get squashed together. Slowly they set hard into rock.

- Metamorphic rock is rock that has been changed. Sometimes an igneous or a sedimentary rock is heated or squashed under the ground, so that it changes. Metamorphic rock is often harder than the rock that it was before it changed. Marble and slate are among the metamorphic rocks that people use to make things.

## Faults

All three kinds of rock can be changed by being heated and squeezed by forces in the earth. When this happens, faults (cracks) may appear in the rock. Geologists can learn a lot about the history of the rock by studying the patterns of the fault lines. Earthquakes are caused when a fault breaks suddenly.

## Soil

Soil is the stuff on the ground made of lots of particles (or tiny pieces). The particles of soil come from rocks that have broken down, and from rotting leaves and animals bodies. Soil covers a lot of the surface of the Earth. Plants of all sorts grow in soil.

## Principles of Stratigraphy

Geologists use some simple ideas which help them to understand the rocks they are studying. The following ideas were worked out in the early days of stratigraphy by people like Nicolaus Steno, James Hutton and William Smith:

1. Understanding the past: Geologist James Hutton said "The present is the key to the past". He meant that the sort of changes that are happening to the Earth's surface now are the same sorts of things that happened in the past. Geologists can understand things that happened millions of years ago, by looking at the changes which are happening today.

2. Horizontal strata: The layers in a sedimentary rock must have been horizontal (flat) when they were deposited (laid down).

3. The age of the strata: Layers at the bottom must be older than layers at the top, unless all the rocks have been turned over.

4. In sedimentary rocks that are made of sand or gravel, the sand or gravel must have come from an older rock.

5.  The age of faults: If there is a crack or fault in a rock, then the fault is younger than the rock. Rocks are in strata (lots of layers). A geologist can see if the faults go through all the layer, or only some. This helps to tell the age of the rocks.

6.  The age of a rock which cuts through other rocks: If an igneous rock cuts across sedimentary layers, it must be younger than the sedimentary rock.

7.  The relative age of fossils: A fossil in one rock type must be about the same age as the same type of fossil in the same type of rock *in a different place*. Likewise, a fossil in a rock layer below must be earlier than one in a higher layer.

The history of geology is concerned with the development of the natural science of geology. Geology is the scientific study of the origin, history, and structure of the Earth.

Scotsman James Hutton is considered to be the father of modern geology

Some of the first geological thoughts were about the origin of the Earth. Ancient Greece developed some primary geological concepts concerning the origin of the Earth. BC Aristotle made critical observations of the slow rate of geological change. He observed the composition of the land and formulated a theory where the Earth changes at a slow rate and that these changes cannot be observed during one person's lifetime. Aristotle developed one of the first evidentially based concepts connected to the geological realm regarding the rate at which the Earth physically changes.

However, it was his successor at the Lyceum, the philosopher Theophrastus, who made the greatest progress in antiquity in his work *On Stones*. He described many minerals and ores both from local mines such as those at Laurium near Athens, and further afield. He also quite naturally discussed types of marble and building materials like limestones, and attempted a primitive classification of the properties of minerals by their properties such as hardness.

Much later in the Roman period, Pliny the Elder produced a very extensive discussion of many more minerals and metals then widely used for practical ends. He was among the first to correctly identify the origin of amber as a fossilized resin from trees by the observation of insects trapped within some pieces. He also laid the basis of crystallography by recognising the octahedral habit of diamond.

A mosquito and a fly in this Baltic amber necklace are between 40 and 60 million years old

The slightly misshapen octahedral shape of this rough diamond crystal in matrix is typical of the mineral. Its lustrous faces also indicate that this crystal is from a primary deposit.

Abu al-Rayhan al-Biruni (AD 973–1048) was one of the earliest Muslim geologists, whose works included the earliest writings on the geology of India, hypothesizing that the Indian subcontinent was once a sea.

Ibn Sina (Avicenna, 981–1037), a Persian polymath, made significant contributions to geology and the natural sciences (which he called *Attabieyat*) along with other natural philosophers such as Ikhwan Al-Safa and many others. Ibn Sina wrote an encyclopaedic work entitled "*Kitab al-Shifa*" (the Book of Cure, Healing or Remedy from ignorance), in which Part 2, contains his commentary on Aristotle's Mineralogy and Meteorology, in six chapters: Formation of mountains, The advantages of mountains in the formation of clouds; Sources of water; Origin of earthquakes; Formation of minerals; The diversity of earth's terrain.

In medieval China, one of the most intriguing naturalists was Shen Kuo (1031–1095), a polymath personality who dabbled in many fields of study in his age. In terms of geology, Shen Kuo is one of the first naturalists to have formulated a theory of geomorphology. This was based on his observations of sedimentary uplift, soil erosion, deposition of silt, and marine fossils found in the Taihang Mountains, located hundreds of miles from the Pacific Ocean. He also formulated a theory of gradual climate change, after his observation of ancient petrified bamboos found in a preserved state underground near Yanzhou (modern Yan'an), in the dry northern climate of Shaanxi province. He formulated a hypothesis for the process of land formation: based on his observation of fossil shells in a geological stratum in a mountain hundreds of miles from the ocean, he inferred that the land was formed by erosion of the mountains and by deposition of silt.

A portrait of Whiston with a diagram demonstrating his theories of cometary catastrophism best described in *A New Theory of the Earth*

It was not until the geology made great strides in its development. At this time, geology became its own entity in the world of natural science. It was discovered by the Christian world that different translations of the Bible contained different versions of the biblical text. The one entity that remained consistent through all of the interpretations was that the Deluge had formed the world's geology and geography. To prove the Bible's authenticity, individuals felt the need to demonstrate with scientific evidence that the Great Flood had in fact occurred. With this enhanced desire for data came an increase in observations of the Earth's composition, which in turn led to the discovery of fossils. Although theories that resulted from the heightened interest in the Earth's composition were often manipulated to support the concept of the Deluge, a genuine outcome was a greater interest in the makeup of the Earth. Due to the strength of Christian, the theory of the origin of the Earth that was most widely accepted was *A New Theory of the Earth* published in 1696, by William Whiston. Whiston used Christian reasoning to "prove" that the Great Flood had occurred and that the flood had formed the rock strata of the Earth.

The heated debate between religion and science over the Earth's origin further propelled interest in the Earth and brought about more systematic identification techniques of the Earth's strata. The Earth's strata can be defined as horizontal layers of rock having approximately the same composition throughout. An important pioneer in the science was Nicolas Steno. Steno was trained in the classical texts on science; however, by 1659 he seriously questioned accepted knowledge of the natural world. Importantly, he questioned the idea that fossils grew in the ground, as well as common explanations of rock formation. His investigations and his subsequent conclusions on these topics have led scholars to consider him one of the founders of modern stratigraphy and geology.

From this increased interest in the nature of the Earth and its origin, came a heightened attention to minerals and other components of the Earth's crust. Moreover, the increasing economic importance of mining in Europe during the mid to late possession of accurate knowledge about ores and their natural distribution vital. Scholars began to study the makeup of the Earth in a systematic manner, with detailed comparisons and descriptions not only of the land itself, but of the semi-precious metals it contained, which had great commercial value. For example, in 1774 Abraham Gottlob Werner published the book *Von den äusserlichen Kennzeichen der Fossilien (On the External Characters of Minerals)*, which brought him widespread recognition because he presented a detailed system for identifying specific minerals based on external characteristics. The more efficiently productive land for mining could be identified and the semi-precious metals could be found, the more money could be made. This drive for economic gain propelled geology into the limelight and made it a popular subject to pursue. With an increased number of people studying it, came more detailed observations and more information about the Earth.

The history of the Earth—namely the divergences between the accepted religious concept and factual evidence—once again became a popular topic for discussion in society. In 1749, the French naturalist Georges-Louis Leclerc, Comte de Buffon published his *Histoire Naturelle,* in which he attacked the popular Biblical accounts given by Whiston and other ecclesiastical theorists of the history of Earth. From experimentation with cooling globes, he found that the age of the Earth was not only 4,000 or 5,500 years as inferred from the Bible, but rather 75,000 years. Another individual who described the history of the Earth with reference to neither God nor the Bible was the philosopher Immanuel Kant, who published his *Universal Natural History and Theory of the Heavens (Allgemeine Naturgeschichte und Theorie des Himmels)* in 1755. From the works of these respected men, as well as others. This questioning represented a turning point in the study of the Earth. It was now possible to study the history of the Earth from a scientific perspective without religious preconceptions.

With the application of scientific methods to the investigation of the Earth's history, the study of geology could become a distinct field of science. To begin with, the terminology and definition of what constituted geological study had to be worked out. The term "geology" was first used technically in publications by two Genevan naturalists, Jean-André Deluc and Horace-Bénédict de Saussure, though "geology" was not well received as a term until it was taken up in the very influential compendium, the *Encyclopédie*, published beginning in 1751 by Denis Diderot. Once the term was established to denote the study of the Earth and its history, geology slowly became more generally recognized as a distinct science that could be taught as a field of study at educational institutions. In 1741 the best-known institution in the field of natural history, the National Museum of Natural History in France, created the first teaching position designated specifically for geology. This was an important step in further promoting knowledge of geology as a science and in recognizing the value of widely disseminating such knowledge.

By the 1770s, chemistry was starting to play a pivotal role in the theoretical foundation of geology and two opposite theories with committed followers emerged. These contrasting theories offered differing explanations of how the rock layers of the Earth's surface had formed. One suggested that a liquid inundation, perhaps like the biblical deluge, had created all geological strata. The theory extended chemical theories that had been developing and was promoted by Scotland's John Walker, Sweden's Johan Gottschalk Wallerius and Germany's Abraham Werner. Of these names, Werner's views become internationally influential around 1800. He argued that the Earth's layers, including basalt and granite, had formed as a precipitate from an ocean that covered the entire Earth. Werner's system was influential and those who accepted his theory were known as Diluvianists or Neptunists. The Neptunist thesis was the most popular during especially for those who were chemically trained. However, another thesis slowly gained currency from the 1780s forward. Instead of water, naturalists such as Buffon had suggested that strata had been formed through heat (or fire). The thesis was modified and expanded by the Scottish naturalist James Hutton during the 1780s. He argued against the theory of Neptunism, proposing instead the theory of based on heat. Those who followed this thesis during the early nineteenth century referred to this view as Plutonism: the formation of the Earth through the gradual solidification of a molten mass at a slow rate by the same processes that had occurred throughout history and continued in the present day. This led him to the conclusion that the Earth was immeasurably old and could not possibly be explained within the limits of the chronology inferred from the Bible. Plutonists believed that volcanic processes were the chief agent in rock formation, not water from a Great Flood.

Bust of William Smith, in the Oxford University Museum of Natural History.

In the early 19th century, the mining industry and Industrial Revolution stimulated the rapid development of the stratigraphic column - "the sequence of rock formations arranged according to their order of formation in time." In England, the mining surveyor William Smith, starting in the 1790s, found empirically that fossils were a highly effective means of distinguishing between otherwise similar formations of the landscape as he travelled the country working on the canal system and produced the first geological map of Britain. At about the same time, the French comparative anatomist Georges Cuvier assisted by his colleague Alexandre Brogniart at the École des Mines de Paris realized that the relative ages of fossils could be determined from a geological standpoint; in terms of what layer of rock the fossils are located and the distance these layers of rock are from the surface of the Earth. Through the synthesis of their findings, Brogniart and Cuvier realized that different strata could be identified by fossil contents and thus each stratum could be assigned to a unique position in a sequence. After the publication of Cuvier and Brongniart's book, "Description Geologiques des Environs de Paris" in 1811, which outlined the concept, stratigraphy became very popular amongst geologists; many hoped to apply this concept to all the rocks of the Earth. During this century various geologists further refined and completed the stratigraphic column. For instance, in 1833 while Adam Sedgwick was mapping rocks that he had established were from the Cambrian Period, Charles Lyell was elsewhere suggesting a subdivision of the Tertiary Period; whilst Roderick Murchison, mapping into Wales from a different direction, was assigning the upper parts of Sedgwick's *Cambrian* to the lower parts of his own Silurian Period. The stratigraphic column was significant because it supplied a method to assign a relative age of these rocks by slotting them into different positions in their stratigraphical sequence. This created a global approach to dating the age of the Earth and allowed for further correlations to be drawn from similarities found in the makeup of the Earth's crust in various countries.

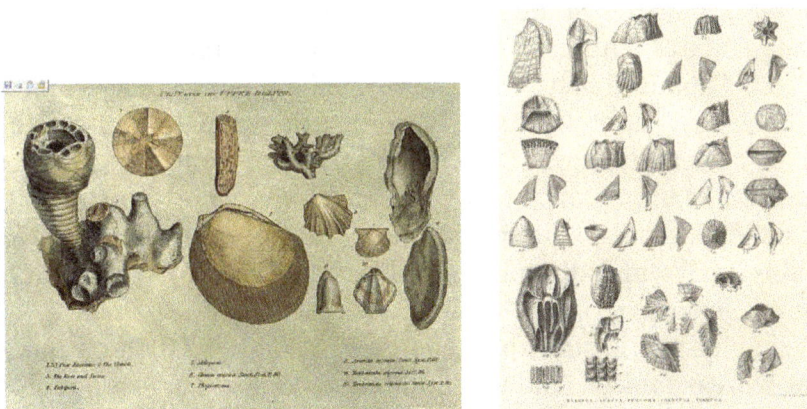

Engraving from William Smith's monograph on identifying strata by fossils

In early nineteenth-century Britain, catastrophism was adapted with the aim of reconciling geological science with religious traditions of the biblical Great Flood. In the early 1820s English geologists including William Buckland and Adam Sedgwick interpreted "diluvial" deposits as the outcome of Noah's flood, but by the end of the decade they revised their opinions in favour of local inundations. Charles Lyell challenged catastrophism with the publication in 1830 of the first volume of his book *Principles of Geology* which presented a variety of geological evidence from England, France, Italy and Spain to prove Hutton's ideas of gradualism correct. He argued that most geological change had been very gradual in human history. Lyell provided evidence for Uniformitarianism; a geological doctrine that processes occur at the same rates in the present as they did in the past and account for all of the Earth's geological features. Lyell's works were popular and widely read, the concept of Uniformitarianism had taken a strong hold in geological society.

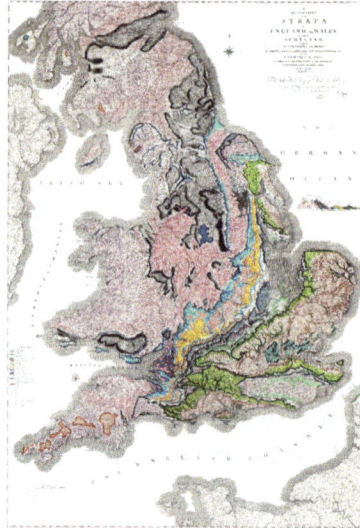

Geological map of Great Britain by William Smith.

During the same time that the stratigraphic column was being completed, imperialism drove several countries in the early to mid 19th century to explore distant lands to expand their empires. This gave naturalists the opportunity to collect data on these voyages. In 1831 Captain Robert FitzRoy, given charge of the coastal survey expedition of HMS *Beagle*, sought a suitable naturalist to examine the land and give geological advice. This fell to Charles Darwin, who had just completed his BA degree and had accompanied Sedgwick on a two-week Welsh mapping expedition after taking his Spring course on geology. Fitzroy gave Darwin Lyell's *Principles of Geology*, and Darwin became Lyell's first disciple, inventively theorising on uniformitarian principles about the geological processes he saw, and challenging some of Lyell's ideas. He speculated about the Earth expanding to explain uplift, then on the basis of the idea that ocean areas sank as land was uplifted, theorised that coral atolls grew from fringing coral reefs round sinking volcanic islands. This idea was confirmed when the *Beagle* surveyed the Cocos (Keeling) Islands, and in 1842 he published his theory on *The Structure and Distribution of Coral Reefs*. Darwin's discovery of giant fossils helped to establish his reputation as a geologist, and his theorising about the causes of their extinction led to his theory of evolution by natural selection published in *On the Origin of Species* in 1859.

Economic motivations for the practical use of geological data caused governments to support geological research. During the 19th century the governments of several countries including Canada, Australia, Great Britain and the United States funded geological surveying that would produce geological maps of vast areas of the countries. Geological surveying provides the location of useful minerals and such information could be used to benefit the country's mining industry. With the government funding of geological research, more individuals could study geology with better technology and techniques, leading to the expansion of the field of geology.

In the 19th century, scientific inquiry had estimated the Age of the Earth in terms of millions of years. By the early 20th Century radiogenic isotopes had been discovered and Radiometric Dating had been developed. In 1911 Arthur Holmes dated a sample from Ceylon at 1.6 billion years old using lead isotopes. In 1921, attendees at the yearly meeting of the British Association for the Advancement of Science came to a rough consensus that the Age of the Earth was a few billion years old, and that radiometric dating was credible. Holmes published The Age of the Earth, an Introduction to Geological Ideas in 1927 in which he presented a range of 1.6 to 3.0 billion years. Subsequent dating has taken the Age of the Earth to around 4.55 billion years. Theories that did not comply with the scientific evidence that established the age of the Earth could no longer be accepted.

## 20th Century

In 1862, the physicist William Thomson, 1st Baron Kelvin, published calculations that fixed the age of Earth at between 20 million and 400 million years. He assumed that Earth had formed as a completely molten object, and determined the amount of time it would take for the near-surface to cool to its present temperature. With the discovery of radioactive decay the age of the Earth was pushed back even further. Arthur Holmes, was a pioneer of geochronology. In 1913 Holmes was on the staff of Imperial College, when he published his famous book *The Age of the Earth* in which he argued strongly in favour of the use of radioactive dating methods rather than methods based on geological sedimentation or cooling of the earth (many people still clung to Lord Kelvin's calculations of less than 100 million years). Holmes estimated the oldest Archean rocks to be 1,600 million years, but did not speculate about the Earth's age. By this time the discovery of isotopes had complicated the calculations and he spent the next years grappling with these. His promotion of the theory over the next decades earned him the nickname of Father of Modern Geochronology. By 1927 he had revised this figure to 3,000 million years and in the 1940s to 4,500±100 million years, based on measurements of the relative abundance of uranium isotopes established by Alfred O. C. Nier. The general method is now known as the Holmes-Houterman model after Fritz Houtermans who published in the same year, 1946. The established age of the Earth has been refined since then but has not significantly changed.

Alfred Wegener

In 1912 Alfred Wegener proposed the theory of Continental Drift. This theory suggests that the shapes of continents and matching coastline geology between some continents indicates they were joined together in the past and formed a single landmass known as Pangaea; thereafter they separated and drifted like rafts over the ocean floor, currently reaching their present position. Additionally, the theory of continental drift offered a possible explanation as to the formation of mountains; Plate Tectonics built on the theory of continental drift.

Unfortunately, Wegener provided no convincing mechanism for this drift, and his ideas were not generally accepted during his lifetime. Arthur Homes accepted Wegener's theory and provided a mechanism: mantle convection, to cause the continents to move. However, it was not until after the Second World War that new evidence started to accumulate that supported continental drift. There followed a period of 20 extremely exciting years where the Theory of Continental Drift developed from being believed by a few to being the cornerstone of modern Geology. Beginning in 1947 research found new evidence about the ocean floor, and in 1960 Bruce C. Heezen published the concept of mid-ocean ridges.Soon after this, Robert S. Dietz and Harry H. Hess proposed that the oceanic crust forms as the seafloor spreads apart along mid-ocean ridges in seafloor spreading. This was seen as confirmation of mantle convection and so the major stumbling block to the theory was removed. Geophysical evidence suggested lateral motion of continents and that oceanic crust is younger than continental crust. This geophysical evidence also spurred the hypothesis of paleomagnetism, the record of the orientation of the Earth's magnetic field recorded in magnetic minerals. British geophysicist S. K. Runcorn suggested the concept of paleomagnetism from his finding that the continents had moved relative to the Earth's magnetic poles. Tuzo Wilson, who was a promoter of the sea floor spreading hypothesis and continental drift from the very beginning, added the concept of transform faults to the model, completing the classes of fault types necessary to make the mobility of the plates on the globe function. A symposium on continental drift was held at the Royal Society of London in 1965 must be regarded as the official start of the acceptance of plate tectonics by the scientific community. The abstracts from the symposium are issued as Blacket, Bullard, Runcorn;1965.In this symposium, Edward Bullard and co-workers showed with a computer calculation how the continents along both sides of the Atlantic would best fit to close the ocean, which became known as the famous "Bullard's Fit". By the late 1960s the weight of the evidence available saw Continental Drift as the generally accepted theory.

## Modern Geology

By applying sound stratigraphic principles to the distribution of craters on the Moon, it can be argued that almost overnight, Gene Shoemaker took the study of the Moon away from Lunar astronomers and gave it to Lunar geologists.

In recent years, geology has continued its tradition as the study of the character and origin of the Earth, its surface features and internal structure. What changed in the later 20th century is the perspective of geological study. Geology was now studied using a more integrative approach, considering the Earth in a broader context encompassing the atmosphere, biosphere and hydrosphere. Satellites located in space that take wide scope photographs of the Earth provide such a perspective. In 1972, The Landsat Program, a series of satellite missions jointly managed by NASA and the U.S. Geological Survey, began supplying satellite images that can be geologically analyzed. These images can be used to map major geological units, recognize and correlate rock types for vast regions and track the movements of Plate Tectonics. A few applications of this data include the ability to produce geologically detailed maps, locate sources of natural energy and predict possible natural disasters caused by plate shifts.

## Geologic Time Scale

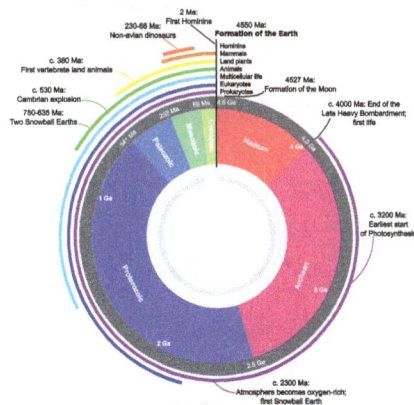

This clock representation shows some of the major units of geological time and definitive events of Earth history.

The Hadean eon represents the time before fossil record of life on Earth; its upper boundary is now regarded as 4.0 Ga (billion years ago). Other subdivisions reflect the evolution of life; the Archean and Proterozoic are both eons, the Palaeozoic, Mesozoic and Cenozoic are eras of the Phanerozoic eon. The three million year Quaternary period, the time of recognizable humans, is too small to be visible at this scale.

The geologic time scale (GTS) is a system of chronological dating that relates geological strata (stratigraphy) to time. It is used by geologists, paleontologists, and other Earth scientists to describe the timing and relationships of events that have occurred during Earth's history. The table of geologic time spans, presented here, agree with the nomenclature, dates and standard color codes set forth by the International Commission on Stratigraphy (ICS).

The primary defined divisions of time are eons, in sequence the Hadean, the Archean, the Proterozoic and the Phanerozoic. The first three of these can be referred to collectively as the Precambrian supereon. Eons are divided into eras, which are in turn divided into periods, epochs and ages.

The following four timelines show the geologic time scale. The first shows the entire time from the formation of the Earth to the present, but this gives little space for the most recent eon. Therefore, the second timeline shows an expanded view of the most recent eon. In a similar way, the most recent era is expanded in the third timeline, and the most recent period is expanded in the fourth timeline.

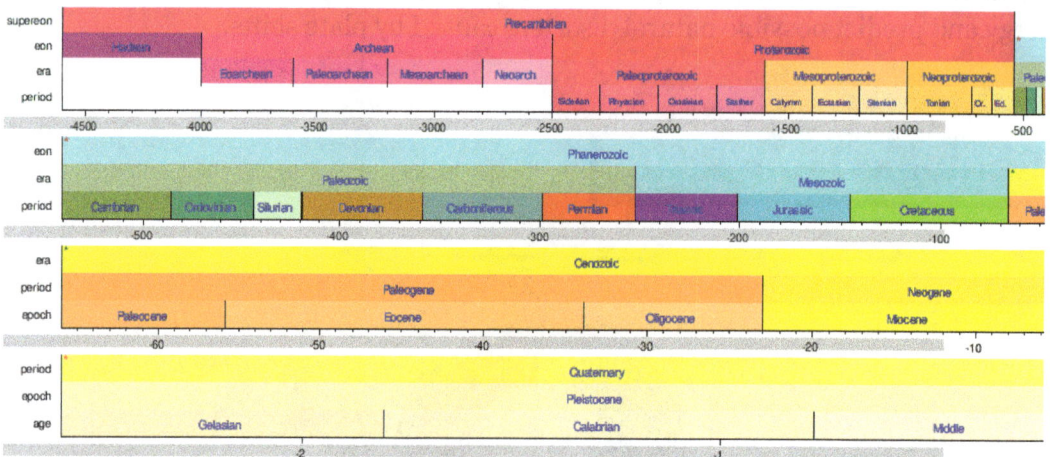

Millions of Years

Corresponding to eons, eras, periods, epochs and ages, the terms "eonothem", "erathem", "system", "series", "stage" are used to refer to the layers of rock that belong to these stretches of geologic time in Earth's history.

Geologists qualify these units as "early", "mid", and "late" when referring to time, and "lower", "middle", and "upper" when referring to the corresponding rocks. For example, the lower Jurassic Series in chronostratigraphy corresponds to the early Jurassic Epoch in geochronology. The adjectives are capitalized when the subdivision is formally recognized, and lower case when not; thus "early Miocene" but "Early Jurassic."

## Rationale

Evidence from radiometric dating indicates that Earth is about 4.54 billion years old. The geology or *deep time* of Earth's past has been organized into various units according to events which took place. Different spans of time on the GTS are usually marked by corresponding changes in the composition of strata which indicate major geological or paleontological events, such as mass extinctions. For example, the boundary between the Cretaceous period and the Paleogene period is defined by the Cretaceous–Paleogene extinction event, which marked the demise of the non-avian dinosaurs and many other groups of life. Older time spans, which predate the reliable fossil record (before the Proterozoic eon), are defined by their absolute age.

Geologic units from the same time but different parts of the world often look different and contain different fossils, so the same time-span was historically given different names in different locales. For example, in North America, the Lower Cambrian is called the Waucoban series that is then subdivided into zones based on succession of trilobites. In East Asia and Siberia, the same unit is split into Alexian, Atdabanian, and Botomian stages. A key aspect of the work of the International Commission on Stratigraphy is to reconcile this conflicting terminology and define universal horizons that can be used around the world.

Some other planets and moons in the Solar System have sufficiently rigid structures to have preserved records of their own histories, for example, Venus, Mars and the Earth's Moon. Dominantly fluid planets, such as the gas giants, do not preserve their history in a comparable manner. Apart from the Late Heavy Bombardment, events on other planets probably had little direct influence on the Earth, and events on Earth had correspondingly little effect on those planets. Construction of a time scale that links the planets is, therefore, of only limited relevance to the Earth's time scale, except in a Solar System context. The existence, timing, and terrestrial effects of the Late Heavy Bombardment are still debated.

In Ancient Greece, Aristotle (384-322 BCE) observed that fossils of seashells in rocks resembled those found on beaches – he inferred that the fossils in rocks were formed by living animals, and he reasoned that the positions of land and sea had changed over long periods of time. Leonardo da Vinci (1452–1519) concurred with Aristotle's interpretation that fossils represented the remains of ancient life.

The Persian geologist Avicenna (Ibn Sina, died 1037) and the Dominican bishop Albertus Magnus (died 1280) extended Aristotle's explanation into a theory of a petrifying fluid. Avicenna also first proposed one of the principles underlying geologic time scales, the law of superposition of strata, while discussing the origins of mountains in *The Book of Healing* (1027). The Chinese naturalist Shen Kuo (1031–1095) also recognized the concept of "deep time".

## Establishment of Primary Principles

In the late Nicholas Steno (1638–1686) pronounced the principles underlying geo-logic (geological) time scales. Steno argued that rock layers (or strata) were laid down in succession, and that each represents a "slice" of time. He also formulated the law of superposition, which states that any given stratum is probably older than those above it and younger than those below it. While Steno's principles were sim-ple, applying them proved challenging. Over the course of the geologists realized that:

1.  Sequences of strata often become eroded, distorted, tilted, or even inverted af-ter deposition

2.  Strata laid down at the same time in different areas could have entirely different appearances

3.  The strata of any given area represented only part of Earth's long history

The Neptunist theories popular at this time proposed that all rocks had precipitated out of a single enormous flood. A major shift in thinking came when James Hutton presented his *Theory of the Earth; or, an Investigation of the Laws Observable in the Composition, Dissolution, and Restoration of Land Upon the Globe* before the Royal Society of Edinburgh in March and April 1785. John McPhee asserts that "as things ap-pear from the perspective of the 20th century, James Hutton in those readings became the founder of modern geology". Hutton proposed that the interior of Earth was hot, and that this heat was the engine which drove the creation of new rock: land was eroded by air and water and deposited as layers in the sea; heat then consolidated the sediment into stone, and uplifted it into new lands. This theory, known as "Plutonism", stood in contrast to the "Neptunist" flood-oriented theory.

## Formulation of Geologic Time Scale

The first serious attempts to formulate a geologic time scale that could be applied anywhere on Earth were made. The most influential of those early attempts (cham-pioned by Werner, among others) divided the rocks of Earth's crust into four types: Primary, Secondary, Tertiary, and Quaternary. Each type of rock, according to the theory, formed during a specific period in Earth history. It was thus possible to speak of a "Tertiary Period" as well as of "Tertiary Rocks." Indeed, "Tertiary" (now Paleo-gene and Neogene) remained in use as the name of a geological period well into the 20th century and "Quaternary" remains in formal use as the name of the current period.

The identification of strata by the fossils they contained, pioneered by William Smith, Georges Cuvier, Jean d'Omalius d'Halloy, and Alexandre Brongniart in the early 19th century, enabled geologists to divide Earth history more precisely. It also enabled them to correlate strata across national (or even continental) boundaries. If two strata (however distant in space or different in composition) contained the same fossils, chances were good that they had been laid down at the same time. Detailed studies between 1820 and 1850 of the strata and fossils of Europe produced the sequence of geologic periods still used today.

## Naming of Geologic Periods, Eras and Epochs

Early work on developing the geologic time scale was dominated by British geologists, and the names of the geologic periods reflect that dominance. The "Cambrian", (the classical name for Wales) and the "Ordovician", and "Silurian", named after ancient Welsh tribes, were periods defined using stratigraphic sequences from Wales. The "Devonian" was named for the English county of Devon, and the name "Carboniferous" was an adaptation of "the Coal Measures", the old British geologists' term for the same set of strata. The "Permian" was named after Perm, Russia, because it was defined using strata in that region by Scottish geologist Roderick Murchison. However, some periods were defined by geologists from other countries. The "Triassic" was named in 1834 by a German geologist Friedrich Von Alberti from the three distinct layers (Latin *trias* meaning triad)—red beds, capped by chalk, followed by black shales—that are found throughout Germany and Northwest Europe, called the 'Trias'. The "Jurassic" was named by a French geologist Alexandre Brongniart for the extensive marine limestone exposures of the Jura Mountains. The "Cretaceous" (from Latin *creta* meaning 'chalk') as a separate period was first defined by Belgian geologist Jean d'Omalius d'Halloy in 1822, using strata in the Paris basin and named for the extensive beds of chalk (calcium carbonate deposited by the shells of marine invertebrates) found in Western Europe.

British geologists were also responsible for the grouping of periods into eras and the subdivision of the Tertiary and Quaternary periods into epochs. In 1841 John Phillips published the first global geologic time scale based on the types of fossils found in each era. Phillips' scale helped standardize the use of terms like *Paleozoic* ("old life") which he extended to cover a larger period than it had in previous usage, and *Mesozoic* ("middle life") which he invented.

## Dating of Time Scales

When William Smith and Sir Charles Lyell first recognized that rock strata represented successive time periods, time scales could be estimated only very imprecisely since estimates of rates of change were uncertain. While creationists had been proposing dates of around six or seven thousand years for the age of Earth based on the Bible, early geologists were suggesting millions of years for geologic periods, and some were even suggesting a virtually infinite age for Earth. Geologists and paleontologists constructed the geologic table based on the relative positions of different strata and fossils, and estimated the time scales based on studying rates of various kinds of weathering, erosion, sedimentation, and lithification. Until the discovery of radioactivity in 1896 and the development of its geological applications through radiometric dating during the first half of the 20th century, the ages of various rock strata and the age of Earth were the subject of considerable debate.

The first geologic time scale that included absolute dates was published in 1913 by the British geologist Arthur Holmes. He greatly furthered the newly created discipline of geochronology and published the world-renowned book *The Age of the Earth* in which he estimated Earth's age to be at least 1.6 billion years.

In 1977, the *Global Commission on Stratigraphy* (now the International Commission on Stratigraphy) began to define global references known as GSSP (Global Boundary Stratotype Sections and Points) for geologic periods and faunal stages. The commission's most recent work is described in the 2004 geologic time scale of Gradstein et al. A UML model for how the timescale is structured, relating it to the GSSP, is also available.

## The Anthropocene

Popular culture and a growing number of scientists use the term "Anthropocene" informally to label the current epoch in which we are living. The term was coined by Paul Crutzen and Eugene Stoermer in 2000 to describe the current time, in which humans have had an enormous impact on the environment. It has evolved to describe an "epoch" starting some time in the past and on the whole defined by anthropogenic carbon emissions and production and consumption of plastic goods that are left in the ground.

Critics of this term say that the term should not be used because it is difficult, if not nearly impossible, to define a specific time when humans started influencing the rock strata—defining the start of an epoch. Others say that humans have not even started to leave their biggest impact on Earth, and therefore the Anthropocene has not even started yet.

The ICS has not officially approved the term as of September 2015, the ICS has not officially approved the term. The Anthropocene Working Group met in Oslo in April 2016 to consolidate evidence supporting the argument for the Anthropocene as a true geologic epoch. Evidence was evaluated and the group voted to recommend "Anthropocene" as the new geological age in August 2016. Should the International Commission on Stratigraphy approve the recommendation, the proposal to adopt the term will have to be ratified by the International Union of Geological Sciences before its formal adoption as part of the geologic time scale.

## Geological Time Scale

Introduction: Our planet Earth, the only known habitable planet in the known Universe, continues to be a dynamic system. Just as we write and the reader reads, the internal and external physical, biological and chemical processes are continuously reshaping the Earth. Geological processes in general are very slow that we can hardly notice during our life span while some are abrupt and catastrophic. What we observe today on the surface of the Earth is a snapshot resulting from complex interplay of many internal processes, and by understanding these processes; the Geologists have extrapolated the information to reveal the story of the Earth from its inception.

Earth as a Jigsaw: How do we understand the story of the Earth? How do we fit these small pieces of this complex jigsaw puzzle to spatially and temporally validate our understanding of the processes today? The reader may appreciate that even to tell the story of the world for the last one hundred years; historians would often refer to the global events such as Pre-World War I, or post World War II or first oil crisis or second oil crisis. Some others may be familiar with the Facebook timeline where a person is choosing to tell his/her story in a reverse chronological way. Similarly, to tell the story of the Earth, geologists look for global events that can have significant relevance and presence across the planet.

Early work on Geological Time Scale: The initial studies that contributed to the understanding of the geological processes were limited in scope to a region or a field; to correlate and construct the story of the Earth has been a very formidable and complex task. It may be important to mention that from the time when interest in this subject appeared the age of the Earth was not known and information on natural processes was limited to unverified hypotheses. The early efforts to construct this story started in 1600's.

One of the earliest attempts to calculate the age of the Earth was 6000 years by Archbishop James Ussher mentioned in Genesis (Annals of the World, A.D. 1658)

Nicolaus Steno, a Danish scientist, pioneered the field of stratigraphy (study of strata). Contemporary geoscientists to date rely on the timeless principles viz., *The principle of original horizontality* and *the principle of superposition,* proposed by Steno to comprehend sedimentary rocks.

William Smith, British surveyor, developed the concept of faunal succession. Smith's *Principle of faunal succession* on the basis of characteristic fossils in different strata helps in identifying strata of the same age in different outcrops at different locations.

Nicolaus Steno in 1667 found shark tooth while studying the sedimentary strata in Tuscany, what were then called "tongue stones" (Figure). The concept promoted by Nicolaus that these were remains of sharks left in the rock was not accepted at that time. William Smith, worked on understanding the distinct rock types with distinct fossils, that helped him create a sequence that he was able to validate in other parts of England as well. This stratigraphic sequencing was one of the earliest attempts at creating a geological order for assigning relative ages to the rock formations.

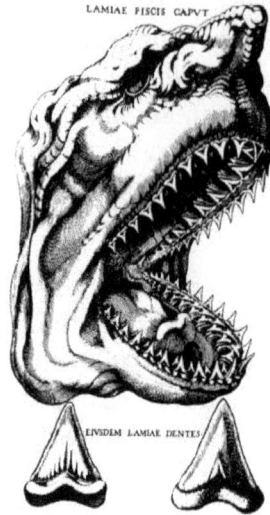

LAMIAE PISCIS CAPVT

EIVSDEM LAMIAE DENTES

Shark tooth drawings made by Nicolaus Steno

Charles Darwin's assumptions on age of the Earth was 6000 years, a common belief during that time, however based on some geology and his own work on natural selection he thought that the age of Earth could be several hundred million years, which was not what Lord Kelvin believed. Lord Kelvin's work based on thermodynamics and Earth's thermal gradient predicted a much younger age.

There were multiple geological evidence pouring in to support a backward revision in the age of the Earth.

At this stage a better understanding on deposition of sedimentary rocks was developed. While observing the then rate of deposition, estimates on time it may have taken to deposit a thick rock succession was made. As an example, the age of thickest sedimentary sequence such as Grand Canyon was placed at 100 Ma, which supported the theory that Earth is much older than the other estimates that were being made.

To simplify our understanding of the story of the Earth, geologists made attempts at the widest divisions of rock strata into eons, periods and epochs. One of the big events that kept shifting the boundary of these divisions was the presence and absence of life, and of simple and complex life.

- Relative ages: Smith's and Steno's principles were applied to outcrops across the globe in the nineteenth century. Similar fossils from formations in different parts of the world were found. Fossil assemblages also matched up for these formations along with other cross cutting relationships. At the end of the century, the geologists could put a time scale together which was the first geologic time scale based on relative ages.

- Absolute age: The story of absolute ages started when the radiometric dating began. Rutherford in 1905 deduced that radioactivity could be used to give an accurate and absolute age for a rock. This started the trend of isotopic dating that could provide absolute ages. However, till about 1953, the age of the Earth was nowhere close to what we know now.

Claire Patterson, American geochemist, calculated age of the Earth to be 4.55 billion years in the year 1956 and this age of earth calculated on the basis of geochronology has remained largely unchanged since then.

Understanding the Modern Geological Time Scale and the basis of its subdivisions:

Modern Geological Time Scale is a result of many centuries of work put together by geologists that includes understanding stratigraphic successions in the field and studying correlations to comprehend relative ages, and determining absolute ages using radiometric dating.

## Intervals of Geologic Time

What is fundamental to the primary divisions of the Geological Time Scale is the presence or absence of fossils and/or distinct set of fossils. The end of a division and start of another demarcates a distinct presence or absence of a particular set of fossils. Eon is the longest subdivision of the Geologic time and two or more Era's form an Eon. Period is the basic unit of Geologic time in which a typical set of rock unit is made, and is usually named after a locality or the distinguishing characteristics of the rocks reported. Devonian, as an example, is named after a locality in England and Carboniferous period for coal bearing sedimentary rocks. However, Palaeogene and Neogene are an exception and mean old and new respectively.

International Commission on Stratigraphy v 2016/04 (Figure) is an agreed upon and elaborate description of the Modern Geological Time Scale, which offers numerical age for each Eon, Era, Period, Epoch and Stage. This represents no finality as numerical ages are subject to further refinement etc.

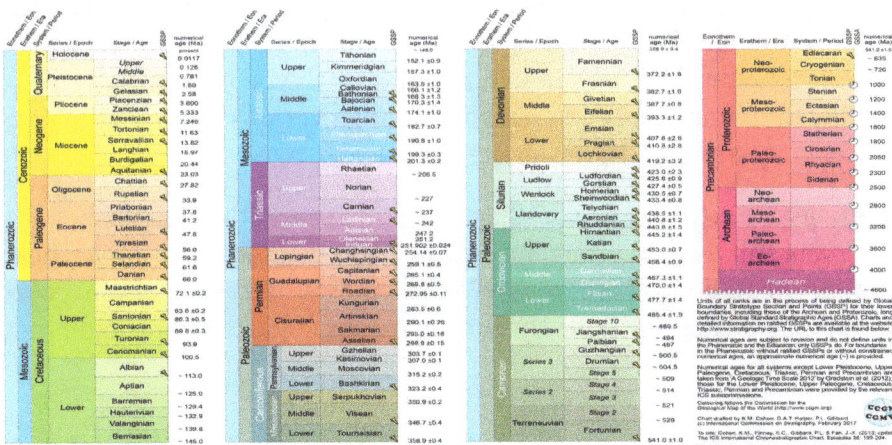

International Chronostratigraphic Chart

Primary Divisions of the Geological Time Scale:

To understand the Geological Time Scale, one can start with its broadest divisions called eons. There are four eons: Hadean, Archean, Proterozoic and Phanerozoic. The Geological Time Scale in its tabular form is presented in a reverse chronological scheme, with the most recent at the top. Hadean (4600 to 4000 Ma) is the oldest eon and is at the base while Phanerozoic the youngest at the top. Hadean: The term 'Hadean' comes from the Greek God of the underworld. This represents a time with absolutely no record of rocks present. One can say that Hadean eon represents those 600 million years of young earth about which we have no records available.

- Archean: Archean eon (4000 to 2500 Ma) was like a new chapter in Earth's history as the Earth is thought to have cooled off sufficiently to allow formation of continents during this eon. Readers may note during the earlier versions of Geological Time Scale that Archean was considered as the oldest (Greek meaning 'beginning). To envision the conditions during Archean is to see the planet as a time of high volcanic activity and so the surviving rocks of the eon are metamorphic and igneous. The readers must also note that owing to the long-time gap between the beginning and end of the period, there must have been considerable variation in the conditions of the Earth at the beginning of the eon and at its end. Archean is now divided into four era's, Eo-Archean, Palaeo-Archean, Meso-Archean and Neo-Archean.

- Eo-Archean (4000 to 3600 Ma) is characterised by earliest crust and greenstone belts in certain shield areas. Palaeo-Archean is younger than Eo-Archean (3600 Ma to 3200 Ma), where the first evidence of fossils ever, dated at 3500 Ma and called stromatolites, were found in Western Australia.

The readers may note that newer scientific discoveries would continue to refine and alter our current understanding of the Geological Time Scale. One such recent discovery of stromatolites from Greenland has taken the stromatolite appearance to 3700 Ma (2016), in the Eo-Archean. Meso-Archean (3600 to 2800 Ma) is characterised by Banded Iron Formations, when oxygen levels on the planet were still low. Some of the earliest super continents formed during this time. Stromatolites diversified during Meso-Archean. Neo-Archean (2800 Ma to 2500 Ma) is when continents started to shape up and continued to fragment and assemble through various supercontinent cycles to acquire present day configuration.

When the Archean began, the Earth's heat flow was nearly three times as high as it is today, and it was still twice the current level at the transition from the Archean to the Proterozoic (2,500 million years ago). The extra heat was the result of a mix of remnant heat from planetary accretion, from the formation of the Earth's core, and produced by radioactive elements.

Although a few mineral grains are known to be Hadean, the oldest rock formations exposed on the surface of the Earth are Archean. Archean rocks are found in Greenland,

Siberia, the Canadian Shield, Montana and Wyoming (exposed parts of the Wyoming Craton), the Baltic Shield, Scotland, India, Brazil, western Australia, and southern Africa. Granitic rocks predominate throughout the crystalline remnants of the surviving Archean crust. Examples include great melt sheets and voluminous plutonic masses of granite, diorite, layered intrusions, anorthosites and monzonites known as sanukitoids. Archean Eon rocks are often heavily metamorphized deep-water sediments, such as graywackes, mudstones, volcanic sediments, and banded iron formations. Volcanic activity was considerably higher than today, with numerous lava eruptions, including unusual types such as komatiite. Carbonate rocks are rare, indicating that the oceans were more acidic due to dissolved carbon dioxide than during the Proterozoic. Greenstone belts are typical Archean formations, consisting of alternating units of metamorphosed mafic igneous and sedimentary rocks. The metamorphosed igneous rocks were derived from volcanic island arcs, while the metamorphosed sediments represent deep-sea sediments eroded from the neighboring island arcs and deposited in a forearc basin. Greenstone belts, being both types of metamorphosed rock, represent sutures between the protocontinents.

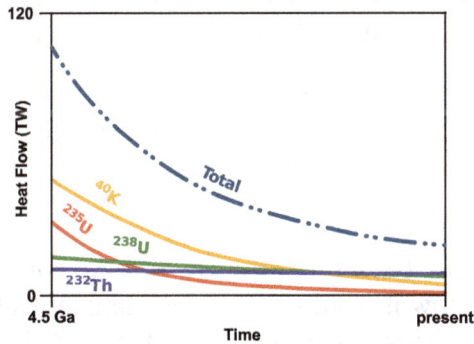

The evolution of Earth's radiogenic heat flow over time

The Earth's continents started to form in the Archean, although details about their formation are still being debated, due to lack of extensive geological evidence. One hypothesis is that rocks that are now in India, western Australia, and southern Africa formed a continent called Ur as of 3,100 Ma. A differing conflicting hypothesis is that rocks from western Australia and southern Africa were assembled in a continent called Vaalbara as far back as 3,600 Ma. Although the first continents formed during this eon, rock of this age makes up only 7% of the present world's cratons; even allowing for erosion and destruction of past formations, evidence suggests that only 5–40% of the present area of continents formed during the Archean.

By the end of the Archaean c. 2500 Ma, plate tectonic activity may have been similar to that of the modern Earth. There are well-preserved sedimentary basins, and evidence of volcanic arcs, intracontinental rifts, continent-continent collisions and widespread globe-spanning orogenic events suggesting the assembly and destruction of one and perhaps several supercontinents. Liquid water was prevalent, and deep oceanic basins are known to have existed attested by the presence of banded iron formations, chert beds, chemical sediments and pillow basalts.

## Early Life

The processes that gave rise to life on Earth are not completely understood, but there is substantial evidence that life came into existence either near the end of the Hadean Eon or early in the Archean Eon.

The earliest evidence for life on Earth are graphite of biogenic origin found in 3.7-billion-year-old metasedimentary rocks discovered in Western Greenland.

The earliest identifiable fossils consist of stromatolites, which are microbial mats formed in shallow water by cyanobacteria. The earliest stromatolites are found in 3.48 billion-year-old sandstone discovered in Western Australia. Stromatolites are found throughout the Archean and become common late in the Archean. Cyanobacteria were instrumental in creating free oxygen in the atmosphere.

Further evidence for early life is found in 3.47-billon-year-old baryte, in the Warrawoona Group of Western Australia. This mineral shows sulfur fractionation of as much as 21.1%, which is evidence of sulfate-reducing bacteria that metabolize sulfur-32 more readily than sulfur-34.

Evidence of life in the Late Hadean is more controversial. In 2015, biogenic carbon has been detected in zircons dated to 4.1 billion years ago, but this evidence is preliminary and needs validation.

Earth was very hostile to life before 4.2–4.3 Ga and the conclusion is that before the Archean Eon, life as we know it would have been challenged by these environmental conditions. While life could have arisen before the Archean, the conditions necessary to sustain life could not have occurred until the Archean Eon.

Life in the Archean was limited to simple single-celled organisms (lacking nuclei), called Prokaryota. In addition to the domain Bacteria, microfossils of the domain Archaea have also been identified. There are no known eukaryotic fossils from the earliest Archean, though they might have evolved during the Archean without leaving any. No fossil evidence has been discovered for ultramicroscopic intracellular replicators such as viruses.

Proterozoic, in Greek means "early life" and the Eon extends from 2500 Ma to 541 Ma; subdivided into three eras, namely Palaeo-Proterozoic (2500 to 1600), Meso-Proterozoic (1600-1000 Ma) and Neo-Proterozoic (1000-541).

Earth's atmosphere changed substantially in the Proterozoic, with increased oxygen levels, referred to as the Great Oxidation Event Although oxygen did show up in the atmosphere in Archean eon, it is only now for the first time that oxygen build up happened to a certain degree that was conducive for many life processes to follow.

Oxygen was a by-product of photosynthesis by cyanobacteria. In its initial stages, whatever oxygen was produced was taken by dissolved iron or organic matter. At one point these sinks became saturated when no more oxygen could be taken up, at which point it started getting released into the atmosphere. It is also argued that free oxygen in the atmosphere may have been detrimental to the anaerobic bacteria of that time, so in that sense, while it paved way for oxygen that is so required for life, it may have caused the first mass extinction. The other thing that this oxygen did was convert methane into carbon dioxide (less potent greenhouse gas and water) which started the Huronian Glaciation, one of the earliest glaciations events in the history of the Earth that led to snowball earth.

In this way, Banded Iron Formations and later red beds coloured by haematite, imply an increase in oxygen levels. The other most significant events in Proterozoic are the formation of the supercontinents and further evolution of life, while stromatolites proliferated, eukaryotes or complex organisms showed up for the first time on the Earth.

Precambrian- this term was used for everything before Cambrian, and if everyone is wondering what is the most often used name Precambrian refers to, it can be called a super eon that represents 'stuff before Cambrian" and combines Hadean, Archean and Proterozoic.

## The Proterozoic Record

The geologic record of the Proterozoic Eon is more complete than that for the preceding Archean Eon. In contrast to the deep-water deposits of the Archean, the Proterozoic features many strata that were laid down in extensive shallow epicontinental seas; furthermore, many of those rocks are less metamorphosed than are Archean ones, and many are unaltered. Studies of these rocks have shown that the eon continued the massive continental accretion that had begun late in the Archean Eon. The Proterozoic Eon also featured the first definitive supercontinent cycles and wholly modern mountain building activity (orogeny).

There is evidence that the first known glaciations occurred during the Proterozoic. The first began shortly after the beginning of the Proterozoic Eon, and evidence of at least four during the Neoproterozoic Era at the end of the Proterozoic Eon, possibly climaxing with the hypothesized Snowball Earth of the Sturtian and Marinoan glaciations.

## The Accumulation of Oxygen

One of the most important events of the Proterozoic was the accumulation of oxygen in the Earth's atmosphere. Though oxygen is believed to have been released by photosynthesis as far back as Archean Eon, it could not build up to any significant degree until mineral sinks of unoxidized sulfur and iron had been filled. Until roughly 2.3 billion years ago, oxygen was probably only 1% to 2% of its current level. The Banded iron formations, which provide most of the world's iron ore, are one mark of that mineral sink process. Their accumulation ceased after 1.9 billion years ago, after the iron in the oceans had all been oxidized.

Red beds, which are colored by hematite, indicate an increase in atmospheric oxygen 2 billion years ago. Such massive iron oxide formations are not found in older rocks. The oxygen buildup was probably due to two factors: a filling of the chemical sinks, and an increase in carbon burial, which sequestered organic compounds that would have otherwise been oxidized by the atmosphere.

## Subduction Processes

The Proterozoic Eon was a very tectonically active period in the Earth's history. The late Archean Eon to Early Proterozoic Eon corresponds to a period of increasing crustal recycling, suggesting subduction. Evidence for this increased subduction activity comes from the abundance of old granites originating mostly after 2.6 Ga. The appearance of eclogites, which metamorphic rocks created by high pressure (>1 GPa), are explained using a model that incorporates subduction. The lack of eclogites that date to the Archean Eon suggests that conditions at that time did not favor the formation of high grade metamorphism and therefore did not achieve the same levels of subduction as was occurring in the Proterozoic Eon. As a result of remelting of basaltic oceanic crust due to subduction, the cores of the first continents grew large enough to withstand the crustal recycling processes. The long-term tectonic stability of those cratons is why we find continental crust ranging up to a few billion years in age. It is believed that 43% of modern continental crust was formed in the Proterozoic, 39% formed in the Archean, and only 18% in the Phanerozoic. Studies by Condie 2000 and Rino et al. 2004 suggest that crust production happened episodically. By isotopically calculating the ages of Proterozoic granitoids it was determined that there were several episodes of rapid increase in continental crust production. The reason for these pulses is unknown, but they seemed to have decreased in magnitude after every period.

## Tectonic History (Supercontinents)

Evidence of collision and rifting between continents raises the question as to what exactly were the movements of the Archean cratons composing Proterozoic continents. Paleomagnetic and geochronological dating mechanisms have allowed the deciphering of Precambrian Supereon tectonics. It is known that tectonic processes of the Proterozoic Eon resemble greatly the evidence of tectonic activity, such as orogenic belts or ophiolite complexes, we see today. Hence, most geologists would conclude that the Earth was active at that time. It is also commonly accepted that during the Precambrian, the Earth went through several supercontinent breakup and rebuilding cycles (Wilson cycle). In the late Proterozoic (most recent), the dominant supercontinent was Rodinia (~1000–750 Ma). It consisted of a series of continents attached to a central craton that forms the core of the North American Continent called Laurentia. An example of an orogeny (mountain building processes) associated with the construction of Rodinia is the Grenville orogeny located in Eastern North America. Rodinia formed after the breakup of the supercontinent Columbia and prior to the assemblage of the supercontinent Gondwana (~500 Ma). The defining orogenic event associated with the formation of Gondwana was the collision of Africa, South America, Antarctica and Australia forming the Pan-African orogeny.

Columbia was dominant in the early-mid Proterozoic and not much is known about continental assemblages before then. There are a few plausible models that explain tectonics of the early Earth pre-Columbia, but the current most plausible theory is that prior to Columbia, there were only a few independent craton formations scattered around the Earth (not necessarily a supercontinent formation like Rodinia or Columbia).

## Life

Stromatolites

South America                              Western Namibia

The first advanced single-celled, eukaryotes and multi-cellular life, Francevillian Group Fossils, roughly coincides with the start of the accumulation of free oxygen. This may have been due to an increase in the oxidized nitrates that eukaryotes use, as opposed to cyanobacteria. It was also during the Proterozoic that the first symbiotic relationships between mitochondria (found in nearly all eukaryotes) and chloroplasts (found in plants and some protists only) and their hosts evolved.

The blossoming of eukaryotes such as acritarchs did not preclude the expansion of cyanobacteria; in fact, stromatolites reached their greatest abundance and diversity during the Proterozoic, peaking roughly 1200 million years ago.

Classically, the boundary between the Proterozoic and the Phanerozoic eons was set at the base of the Cambrian Period when the first fossils of animals including trilobites and archeocyathids appeared. In the second half of the 20th century, a number of fossil forms have been found in Proterozoic rocks, but the upper boundary of the Proterozoic has remained fixed at the base of the Cambrian, which is currently placed at 541 Ma.

Phanerozoic is the current geologic eon. Phanerozoic in Greek means visible life, and it represents the last 541 Ma on the Earth, from the start of the hard-shelled organisms to the dominance of humans. The 541 Ma years have the most records in terms of life on the Earth and other expressions of geological processes.

Phanerozoic eon is divided into Palaeozoic, Mesozoic and Cenozoic eras. Palaeozoic started with the appearance of first complex organisms and it is divided into Cambrian, Ordovician, Silurian, Devonian, Carboniferous and Permian. Cambrian (541 to 485 Ma) which symbolizes the 'explosion of life' is known for proliferation of trilobites. Ordovician (485 to 440 Ma) was the time of primitive fishes. Tectonically, Gondwana the large southern amalgam of continental blocks was at the south pole in the Ordovician period. Glaciation in this period led to mass extinctions. Silurian (440 to 415 Ma) is the age when vascular plants as well as arthropods, fish, both jawed and jawless, became widespread. Devonian (415 Ma to 360 Ma) can be called the 'the age of fishes' first tetrapods, trees, seeds and amphibian evolution. At the end of Devonian another mass extinction happened. Carboniferous (360 and 300 Ma) started with high temperature conditions, trees created carbon that got buried as coal deposits. On the other hand amphibians were dominant. Later carboniferous underwent a period of glaciation. Permian (300 to 250 Ma) is the time when all continents cametogether and formed Pangaea supercontinent and the ocean surrounded it is called Panthalassa. Conifers also evolved during this period and this period ended with mass extinctions.

## Proterozoic-Phanerozoic Boundary

The Proterozoic-Phanerozoic boundary is at 541 million years ago. In the 19th century, the boundary was set at time of appearance of the first abundant animal (metazoan) fossils but several hundred groups (taxa) of metazoa of the earlier Proterozoic era have been identified since the systematic study of those forms started in the 1950s. Most geologists and paleontologists would probably set the Proterozoic-Phanerozoic boundary either at the classic point where the first trilobites and reef-building animals (archaeocyatha) such as corals and others appear; at the first appearance of a complex feeding burrow called *Treptichnus pedum*; or at the first appearance of a group of small, generally disarticulated, armored forms termed 'the small shelly fauna'. The three different dividing points are within a few million years of each other.

In the older literature, the term *Phanerozoic* is generally used as a label for the time period of interest to paleontologists, but that use of the term seems to be falling into disuse in more modern literature.

## Eras of the Phanerozoic

The Phanerozoic is divided into three eras: the Paleozoic, Mesozoic, and Cenozoic, which are further subdivided into 12 periods. The Paleozoic features the rise of fish, amphibians and reptiles. The Mesozoic is ruled by the reptiles, and features the evolution of mammals, birds and more famously, dinosaurs. The Cenozoic is the time of the mammals, and more recently, humans.

## Paleozoic Era

The Paleozoic is a time in Earth's history when complex life forms evolved, took their first breath of oxygen on dry land, and when the forerunners of all life on Earth began to diversify. There are six periods in the Paleozoic era: Cambrian, Ordovician, Silurian, Devonian, Carboniferous and Permian.

## Cambrian Period

Trilobites

The Cambrian is the first period of the Paleozoic Era and starts from 541 to 485 million years ago. The Cambrian sparked a rapid expansion in evolution in an event known as the Cambrian Explosion during which the greatest number of creatures evolved in a single period in the history of Earth. Plants like algae evolved, and the fauna was dominated by armored arthropods, such as trilobites. Almost all marine phyla evolved in this period. During this time, the super-continent Pannotia began to break up, most of which later recombined into the super-continent Gondwana.

## Ordovician Period

Cephalaspis, a jawless fish

The Ordovician spans from 485 million years to 440 million years ago. The Ordovician was a time in Earth's history in which many species still prevalent today evolved, such as primitive fish, cephalopods, and coral. The most common forms of life, however, were trilobites, snails and shellfish. More importantly, the first arthropods crept ashore to colonize Gondwana, a continent empty of animal life. By the end of the Ordovician, Gondwana had moved from the equator to the South Pole, and Laurentia had collided with Baltica, closing the Iapetus Ocean. The glaciation of Gondwana resulted in a major drop in sea level, killing off all life that had established along its coast. Glaciation caused a snowball Earth, leading to the Ordovician-Silurian extinction, during which 60% of marine invertebrates and 25% of families became extinct. This is considered the first mass extinction and the second deadliest in the history of Earth.

## Silurian Period

The Silurian spans from 440 million years to 415 million years ago, which saw a warming from Snowball Earth. This period saw the mass evolution of fish, as jaw-less fish became more numerous, jawed fish evolved, and the first freshwater fish evolved, though arthropods, such as sea scorpions, remained the apex predators. Fully terrestrial life evolved, which included early arachnids, fungi, and centipedes. The evolution of vascular plants (Cooksonia) allowed plants to gain a foothold on land. These early terrestrial plants are the forerunners of all plant life on land. During this time, there were four continents: Gondwana (Africa, South America, Australia, Antarctica, India), Laurentia (North America with parts of Europe), Baltica (the rest of Europe), and Siberia (Northern Asia). The recent rise in sea levels provided new habitats for many new species.

## Devonian Period

The Devonian spans from 415 million years to 360 million years ago. Also known as the "Age of the Fish", the Devonian features a huge diversification in fish, including armored fish like *Dunkleosteus* and lobe-finned fish which eventually evolved into the first tetrapods. On land, plant groups diversified incredibly in an event known as the Devonian Explosion during which the first trees evolved, as well as seeds. This event also allowed the diversification of arthropod life as they took advantage of the new habitat. The first amphibians also evolved, and the fish were now at the top of the food chain. Near the end of the Devonian, 70% of all species became extinct in an event known as the Late Devonian extinction, which is the second mass extinction known to have happened.

*Eogyrinus* (an amphibian) of the Carboniferous

# Carboniferous Period

Dimetrodon

The Carboniferous spans from 360 million to 300 million years ago. During this period, average global temperatures were exceedingly high: the early Carboniferous averaged at about 20 degrees Celsius (but cooled to 10 degrees during the Middle Carboniferous). Tropical swamps dominated the Earth, and the large amounts of trees created much of the carbon that became coal deposits (hence the name Carboniferous). The high oxygen levels caused by these swamps allowed massive arthropods, normally limited in size by their respiratory systems, to proliferate. Perhaps the most important evolutionary development of the time was the evolution of amniotic eggs, which allowed amphibians to move farther inland and remain the dominant vertebrates throughout the period. Also, the first reptiles and synapsids evolved in the swamps. Throughout the Carboniferous, there was a cooling pattern, which eventually led to the glaciation of Gondwana as much of it was situated around the south pole, in an event known as the Permo-Carboniferous glaciation or the Carboniferous Rainforest Collapse.

# Permian Period

The Permian spans from 300 million to 250 million years ago and was the last period of the Paleozoic Era. At its beginning, all continents came together to form the super-continent Pangaea, surrounded by one ocean called Panthalassa. The Earth was very dry during this time, with harsh seasons, as the climate of the interior of Pangaea wasn't regulated by large bodies of water. Reptiles and synapsids flourished in the new dry climate. Creatures such as *Dimetrodon* and *Edaphosaurus* ruled the new continent. The first conifers evolved, then dominated the terrestrial landscape. Nearing the end of the period, *Scutosaurus* and gorgonopsids filled the empty desert. Eventually, they disappeared, along with 95% of all life on Earth in an event simply known as "the Great Dying", the world's third mass extinction event and the largest in its history.

# Mesozoic Era

The Mesozoic ranges from 252 million to 66 million years ago. Also known as "the Age of the dinosaurs", the Mesozoic features the rise of reptiles on their 150 million year conquest of the Earth on the land, in the seas, and in the air. There are three periods in the Mesozoic: Triassic, Jurassic, and Cretaceous.

## Triassic Period

The Triassic ranges from 250 million to 200 million years ago. The Triassic is a desolate transitional time in Earth's history between the Permian Extinction and the lush Jurassic Period. It has three major epochs: Early Triassic, Middle Triassic and Late Triassic.

The Early Triassic lasted between 250 million to 247 million years ago, and was dominated by deserts as Pangaea had not yet broken up, thus the interior was arid. The Earth had just witnessed a massive die-off in which 95% of all life became extinct. The most common life on Earth were *Lystrosaurus*, labyrinthodonts, and *Euparkeria* along with many other creatures that managed to survive the Great Dying. Temnospondyli evolved during this time and would be the dominant predator for much of the Triassic.

*Plateosaurus* (a prosauropod)

The Middle Triassic spans from 247 million to 237 million years ago. The Middle Triassic featured the beginnings of the breakup of Pangaea, and the beginning of the Tethys Sea. The ecosystem had recovered from the devastation of the Great Dying. Phytoplankton, coral, and crustaceans all had recovered, and the reptiles began increasing in size. New aquatic reptiles, such as ichthyosaurs and nothosaurs, evolved. Meanwhile, on land, pine forests flourished, as well as mosquitoes and fruit flies. The first ancient crocodilians evolved, which sparked competition with the large amphibians that had long ruled the freshwater world.

The Late Triassic spans from 237 million to 200 million years ago. Following the bloom of the Middle Triassic, the Late Triassic featured frequent rises of temperature, as well as moderate precipitation (10-20 inches per year). The recent warming led to a boom of reptilian evolution on land as the first true dinosaurs evolved, as well as pterosaurs. The climactic change, however, resulted in a large die-out known as the Triassic-Jurassic extinction event, in which all archosaurs (excluding ancient crocodiles), synapsids, and almost all large amphibians became extinct, as well as 34% of marine life in the fourth mass extinction event. The extinction's cause is debated.

# Jurassic Period

Rhamphorhynchus

The Jurassic ranges from 200 million to 145 million years ago, and features three major epochs: Early Jurassic, Middle Jurassic, and Late Jurassic.

The Early Jurassic Epoch spans from 200 million to 175 million years ago. The climate was much more humid than the Triassic, and as a result, the world was very tropical. In the oceans, plesiosaurs, ichthyosaurs and ammonites dominated the seas. On land, dinosaurs and other reptiles dominated the land, with species such as *Dilophosaurus* at the apex. The first true crocodiles evolved, pushing the large amphibians to near extinction. The reptiles rose to rule the world. Meanwhile, the first true mammals evolved, but never exceeded the height of a shrew.

The Middle Jurassic Epoch spans from 175 million to 163 million years ago. During this epoch, reptiles flourished as huge herds of sauropods, such as *Brachiosaurus* and *Diplodicus*, filled the fern prairies of the Middle Jurassic. Many other predators rose as well, such as *Allosaurus*. Conifer forests made up a large portion of the world's forests. In the oceans, plesiosaurs were quite common, and ichthyosaurs were flourishing. This epoch was the peak of the reptiles.

Artist's depiction of a Stegosaurus (inaccurately portrayed with a dragging tail).

The Late Jurassic Epoch spans from 163 million to 145 million years ago. The Late Jurassic featured a massive extinction of sauropods and ichthyosaurs due to the separation of Pangaea into Laurasia and Gondwana in an extinction known as the Jurassic-Cretaceous extinction. Sea levels rose, destroying fern prairies and creating shallows. Ichthyosaurs became extinct whereas sauropods, as a whole, did not; in fact, some species, like *Titanosaurus*, lived until the K-T extinction. The increase in sea-levels opened up the Atlantic sea way which would continue to get larger over time. The divided world would give opportunity for the diversification of new dinosaurs.

## Cretaceous Period

The Cretaceous is the longest period in the Mesozoic, spans from 145 million to 66 million years ago, and is divided into two epochs: Early Cretaceous, and Late Cretaceous.

Tylosaurus (a mosasaur) hunting Xiphactinus

The Early Cretaceous Epoch spans from 145 million to 100 million years ago. The Early Cretaceous saw the expansion of seaways, and as a result, the decline and extinction of sauropods (except in South America). Many coastal shallows were created, and that caused ichthyosaurs to die out. Mosasaurs evolved to replace them as apex species of the seas. Some island-hopping dinosaurs, like *Eustreptospondylus*, evolved to cope with the coastal shallows and small islands of ancient Europe. Other dinosaurs, such as *Carcharodontosaurus* and *Spinosaurus*, rose to fill the empty space that the Jurassic-Cretaceous extinction had created. Of the most successful would be the *Iguanodon* which spread to every continent. Seasons came back into effect and the poles grew seasonally colder. Dinosaurs such as the *Leaellynasaura* inhabited the polar forests year-round, while many dinosaurs, such as the *Muttaburrasaurus*, migrated there during summer . Since it was too cold for crocodiles, it was the last stronghold for large amphibians, such as the *Koolasuchus*. Pterosaurs grew larger as species like *Tapejara* and *Ornithocheirus* evolved. More importantly, the first true birds evolved sparking competition between them and the pterosaurs.

The Late Cretaceous Epoch spans from 100 million to 65 million years ago. The Late Cretaceous featured a cooling trend that would continue into the Cenozoic Era. Eventually, tropical ecology was restricted to the equator and areas beyond the tropic lines featured extreme seasonal changes of weather. Dinosaurs still thrived as new species such as *Tyrannosaurus*, *Ankylosaurus*, *Triceratops* and Hadrosaurs dominated the food web. Pterosaurs, however, were going into a decline as birds took to the skies.

The last pterosaur to die off was *Quetzalcoatlus*. Marsupials evolved within the large conifer forests as scavengers. In the oceans, Mosasaurs ruled the seas to fill the role of the ichthyosaurs, and huge plesiosaurs, such as *Elasmosaurus*, evolved. Also, the first flowering plants evolved. At the end of the Cretaceous, the Deccan Traps and other volcanic eruptions were poisoning the atmosphere. As this was continued, it is thought that a large meteor smashed into Earth, creating the Chicxulub Crater creating the event known as the K-T Extinction, the fifth and most recent mass extinction event, during which 75% of life on Earth became extinct, including all non-avian dinosaurs. Every living thing with a body mass over 10 kilograms became extinct, and the age of the dinosaurs came to an end.

## Cenozoic Era

The Cenozoic featured the rise of mammals as the dominant class of animals, as the end of the age of the dinosaurs left significant evolutionary vacuums. There are three divisions of the Cenozoic: Paleogene, Neogene and Quaternary.

## Paleogene Period

*Basilosaurus* (a whale, despite the name)

The Paleogene spans from the extinction of the dinosaurs, some 66 million years ago, to the dawn of the Neogene 23 million years ago. It features three epochs: Paleocene, Eocene and Oligocene.

The Paleocene Epoch began with the K-T extinction event caused by the impact of a metorite in the area of present-day Yucatan Peninsula and caused the destruction of 75% of all species on Earth. The Early Paleocene saw the recovery of the Earth from that event. The continents began to take their modern shape, but all continents (and India) were separated from each other. Afro-Eurasia was separated by the Tethys Sea, and the Americas were separated by the strait of Panama, as the Isthmus of Panama had not yet formed. This epoch featured a general warming trend, and jungles eventually reached the poles. The oceans were dominated by sharks as the large reptiles that had once ruled became extinct. Archaic mammals, such as creodonts and early primates that evolved during the Mesozoic filled the world. During this time there were no land creatures over 10 kilograms. Mammals were still quite small.

The Eocene Epoch ranged from 56 million to 34 million years ago. In the early Eocene, land animals were small and living in cramped jungles, much like the Paleocene. None had a mass over 10 kilograms. Among them were early primates, whales and horses along with many other early forms of mammals. At the top of the food chains were huge birds, such as *Gastornis*. It is the only time in recorded history that birds ruled the world (excluding their ancestors, the dinosaurs). The temperature was 30 degrees Celsius with little temperature gradient from pole to pole. In the Middle Eocene Epoch, the circum-Antarctic current between Australia and Antarctica formed which disrupted ocean currents worldwide, resulting in global cooling, and caused the jungles to shrink. This allowed mammals to grow; some such as whales to mammoth proportions, which were, by now, almost fully aquatic. Mammals like *Andrewsarchus* were now at the top of the food-chain and sharks were replaced by *Basilosaurus*, whales, as rulers of the seas. The late Eocene Epoch saw the rebirth of seasons, which caused the expansion of savanna-like areas, along with the evolution of grass.

The Oligocene Epoch spans from 33 million to 23 million years ago. The Oligocene featured the expansion of grass which had led to many new species to take advantage, including the first elephants, cats, dogs, marsupials and many other species still prevalent today. Many other species of plants evolved during this epoch also, such as the evergreen trees. The long term cooling continued and seasonal rains patterns established. Mammals continued to grow larger. *Paraceratherium*, the largest land mammal to ever live evolved during this epoch, along with many other perissodactyls in an event known as the Eocene–Oligocene extinction event (Grand Coupure).

## Neogene Period

The Neogene spans from 23.03 million to 2.58 million years ago. It features 2 epochs: the Miocene, and the Pliocene.

The Miocene spans from 23.03 to 5.333 million years ago and is a period in which grass spread further across, effectively dominating a large portion of the world, diminishing forests in the process. Kelp forests evolved, leading to the evolution of new species, such as sea otters. During this time, perissodactyla thrived, and evolved into many different varieties. Alongside them were the apes, which evolved into a 30 species. Overall, arid and mountainous land dominated most of the world, as did grazers. The Tethys Sea finally closed with the creation of the Arabian Peninsula and in its wake left the Black, Red, Mediterranean and Caspian Seas. This only increased aridity. Many new plants evolved, and 95% of modern seed plants evolved in the mid-Miocene.

Animals of the Miocene (*Chalicotherium, Hyenadon,* entelodont)

The Pliocene lasted from 5.333 to 2.58 million years ago. The Pliocene featured dramatic climactic changes, which ultimately led to modern species and plants. The Mediterranean Sea dried up for several million years. Along with these major geological events, *Australopithecus* evolved in Africa, beginning the human branch. The isthmus of Panama formed, and animals migrated between North and South America, wreaking havoc on the local ecology. Climactic changes brought savannas that are still continuing to spread across the world, Indian monsoons, deserts in East Asia, and the beginnings of the Sahara desert. The Earth's continents and seas moved into their present shapes. The world map has not changed much since, save for changes brought about by the glaciations of the Quaternary, such as the Great Lakes.

## Quaternary Period

Megafauna of the Pleistocene (mammoths, cave lions, woolly rhino, reindeer, horses)

The Quaternary spans from 2.58 million years ago to present day, and is the shortest geological period in the Phanerozoic Eon. It features modern animals, and dramatic changes in the climate. It is divided into two epochs: the Pleistocene and the Holocene.

The Pleistocene lasted from 2.58 million to 11,700 years ago. This epoch was marked by ice ages as a result of the cooling trend that started in the Mid-Eocene. There were at least four separate glaciation periods marked by the advance of ice caps as far south as 40 degrees N latitude in mountainous areas. Meanwhile, Africa experienced a trend of desiccation which resulted in the creation of the Sahara, Namib, and Kalahari deserts. Many animals evolved including mammoths, giant ground sloths, dire wolves, saber-toothed cats, and most famously *Homo sapiens*. 100,000 years ago marked the end of one of the worst droughts of Africa, and led to the expansion of primitive man. As the Pleistocene drew to a close, a major extinction wiped out much of the world's megafauna, including some of the hominid species, such as Neanderthals. All the continents were affected, but Africa to a lesser extent. That continent retains many large animals, such as hippos.

The Holocene began 11,700 years ago and lasts until to present day. All recorded history and "the history of the world" lies within the boundaries of the Holocene epoch. Human activity is blamed for a mass extinction that began roughly 10,000 years ago, though the species becoming extinct have only been recorded since the Industrial Revolution. This is sometimes referred to as the "Sixth Extinction". More than 322 species have become extinct due to human activity since the Industrial Revolution.

## Biodiversity

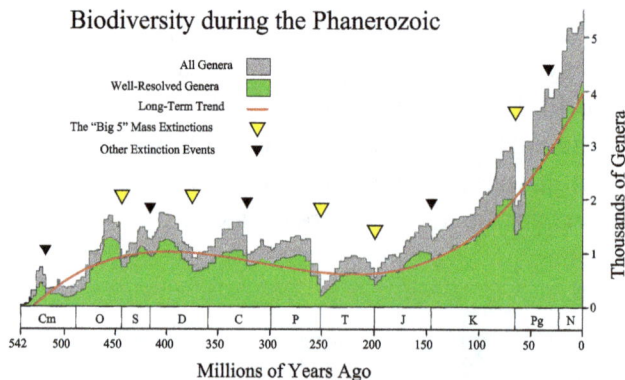

During the Phanerozoic, biodiversity shows a steady but not monotonic increase from near zero to several thousands of genera.

It has been demonstrated that changes in biodiversity through the Phanerozoic correlate much better with the hyperbolic model (widely used in demography and macrosociology) than with exponential and logistic models (traditionally used in population biology and extensively applied to fossil biodiversity as well). The latter models imply that changes in diversity are guided by a first-order positive feedback (more ancestors, more descendants) or a negative feedback that arises from resource limitation, or both. The hyperbolic model implies a second-order positive feedback. The hyperbolic pattern of the human population growth arises from a second-order positive feedback, caused by the interaction of the population size and the rate of technological growth. The character of biodiversity growth in the Phanerozoic Eon can be similarly accounted for by a feedback between the diversity and community structure complexity. It is suggested that the similarity between the curves of biodiversity and human population probably comes from the fact that both are derived from the superposition on the hyperbolic trend of cyclical and random dynamics.

# Geological Processes

Geological processes form mountains in place of oceans and vice versa and how earth scientists unearth the Earth's story.

> "They look at mud and see mountains, in mountains oceans, in oceans mountains to be. They go up to some rock and figure out a story, another rock, another story, and as the stories compile through time they connect-and long case histories are constructed and written from interpreted patterns of data and clues. This is detective work on a scale unimaginable to most detectives, with the notable exception of Sherlock Holmes"
>
> John McPhee, Annals of the Former World

Geological processes and their role in understanding Earth

Geological features that adorn the Earth's surface, both on continents and oceans, in the present form are result of dynamic geological processes that have been operative inside the Earth since the birth of this unique and the only life sustaining planet of the Solar System, almost since last 4.56 billion years. Geological features and events on the Earth are the outcome of a complex interplay of internal and external processes; the former is driven by the internal heat and the latter by the solar heat and atmosphere.

> James Hutton, Scottish geologist, gave the concept of 'Principle of uniformitarianism'.

> Principle of uniformitarianism: The geological processes we see in action on Earth today have analogy to the geological processes which operated throughout the geologic past. This important concept is known as the 'principle of uniformitarianism'.

> Charles Lyell, British geologist, in his book *'Principles of Geology, being an attempt to explain the former changes of the Earth's surface, by reference to causes now in operation'* gave the famous punch-line "The present is the key to the past."

Concept of Geological processes *vis-à-vis* Geological Time Scale

The concept of geological time as discussed in detail in the above section is the greatest geologic discovery that has revolutionized the way earth scientists look at the dynamic Earth in the contemporary times. Pioneering geological field and laboratory observations of scientists such as James Hutton, Charles Lyell, Nicolas Steno, William Smith, Alexander von Humboldt, Alfred Wegner, Claire Patterson to name a few, logically convinced us that the present Erath is an outcome of the past geological processes which operated on multiple time scales since 4.56 billion years. The temporal variation is on a large-to- small scale i.e. million of years to few years to a fraction of seconds! These geological processes are still operative and will continue to operate in future as well. Thus the face of Earth is ever changing and evolving *vis-à-vis* the geological times-cale. Geoscientists have built up an understanding of the present Earth and how it has shaped up and evolved through geologic time.

Lithosphere: Rigid outermost part of the Earth comprising the crust and upper mantle, resting on the weak and ductile asthenosphere.

Plate tectonics: Theory which got acclaim during 1960's on the basis of sea floor spreading. The Earth's surface has been divided into large and small lithospheric plates .The plates are in motion and their relative motion with respect to each other determines the type of boundary: convergent, divergent or transform.

Continental drift: Alfred Wegener, a German meteorologist and explorer, formally proposed the breakup and drifting of continents. His theory was based on his observations of rock types, coastline fit, geological structures and fossils on both the sides of the Atlantic Ocean. In his book entitled '*The Origin of Continents and Oceans*' he proposed that the landmasses of the Earth almost fit together like a jigsaw puzzle

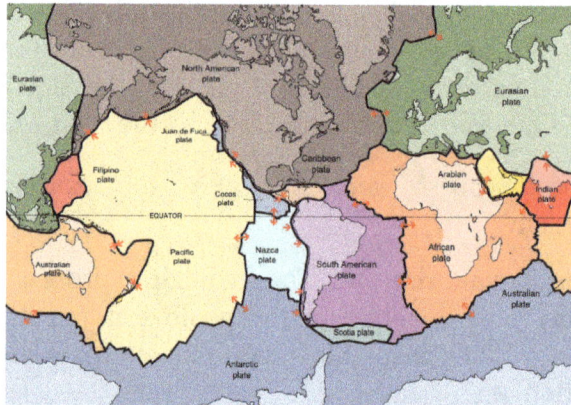

The lithospheric plates of the Earth.

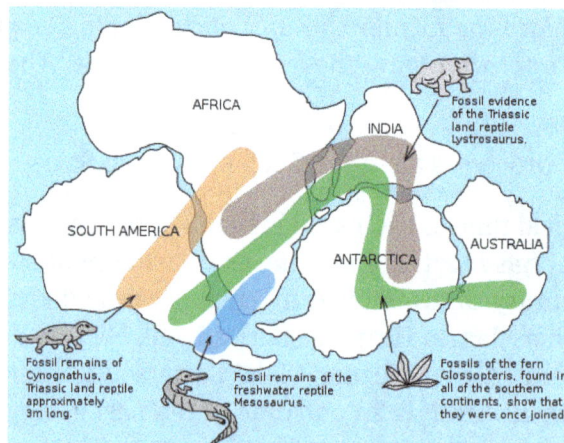

Concept of continental drift.

The principle of uniformitarianism does not imply that all geologic phenomena proceed at the same gradual pace but it gives us clues to understand the basic idea how the Earth has evolved since its inception. Geologic processes can occur in a matter of seconds (volcanic eruptions e.g. Colima volcano, earthquakes e.g. Ring of fire), time span of human life i.e. few years (movement of plates e.g. Rising of Himalaya up to 5 mm every year, migration of sand dunes etc.) or take millions of years (Plate tectonics: Assembly and breaking up of supercontinents e.g. formation of the Himalaya; formation of sedimentary rocks; Figures) to shape the Earth's exterior. Let us consider examples of two geological processes operative on short and long time scales. The example of a short time scale is of drop in altitude of a segment of the Himalaya post-Nepal earthquake, which occurred on 25th April, 2015. "The primary stretch that had its height dropped is a 80-100 km stretch of the Langtang Himal (to the northwest of the capital, Kathmandu)," said (Richard Briggs, a research geologist with the United States Geological Survey (USGS).

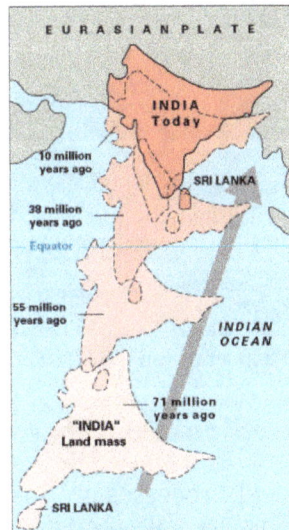

Formation of Himalaya.

The example of a geological process on longer time scale is the formation of the Himalaya (Figure). John McPhee has beautifully summarised the formation of Himalaya in place of the Tethys ocean by movement of plates (a geological process on a large time scale) and how once the marine fossilferrous limestone which formed in the Tethys ocean now constitutes the summit of Mount Everest :

"The Himalayas are the crowning achievement of the Indo-Australian plate. India in the Oligocene crashed head on into Tibet, hit so hard that it not only folded and buckled the plate boundaries but also ploughed into the newly created Tibetan plateau and drove the Himalayas five and a half miles into the sky. The mountains are in some trouble. India has not stopped pushing them, and they are still going up. Their height and volume are already so great they are beginning to melt in their own self-generated radioactive heat. When the climbers in

1953 planted their flags on the highest mountain, they set them in snow over the skeletons of creatures that had lived in a warm clear ocean that India, moving north, blanked out. Possibly as much as 20,000 feet below the sea floor, the skeletal remains had turned into rock. This one fact is a treatise in itself on the movements of the surface of the earth. If by some fiat, I had to restrict all this writing to one sentence; this is the one I would choose: the summit of Mount Everest is marine limestone."

John McPhee, Annals of the Former World

The best way to exemplify John McPhee's quote is to look at the summit of Mt. Everest the highest peak of the world, which is made up of The Qomolangma Formation (Ordovician-limestone, 488-443 million years old) in which many marine fossils such as trilobites, crinoids, and ostracods are entombed (Figure).

Rock exposures on the Mt. Everest

## Internal and External Geological Processes

The internal and external geological processes are instrumental in keeping the Earth dynamic and geologically vibrant. These geological processes manifest on the surface in the form of volcanoes, earthquakes, drifting, movement of lithospheric plates, formation of supercontinents and their dispersal, sedimentary rocks, ore deposits, etc. (Figure). To understand these geological processes geologists track the information concealed in various rocks. There are limitations to the accumulation of this data because of the dynamic nature of the Earth. It is not possible to retrieve all the records preserved in the rocks in the past, and records are obliterated due to weathering and erosion of rocks and subduction of earlier rocks both in oceanic and continental environments. Almost all oceanic crust older than 180 million years has been subducted back into the mantle. To understand the geological story of Earth it is important to have an understanding of the pace of these processes and their driving mechanisms.

There are two principal sources of heat responsible for the internal geological processes:

1. Primordial heat - this heat has been escaping the Earth since its inception, resulting

from the collision of the planetesimals and the later accretion and gravitational differentiation which led to three distinct layers of Earth viz. crust, mantle and core.

2. Radioactive elements - the heat generated by decay of radioactive elements contribute significantly to the Earth's internal heat engine. The internal heat is responsible for generation of melts and convection currents in the mantle which are important in terms of large scale magmatism and plate motions.

Geothermal gradient is the variation in temperature with depth, an indicator of the Earth's internal heat. The average geothermal gradient on the continents is 30° C/km. The geothermal gradient is enhanced in active volcanic areas, mid oceanic ridges and lower in old cratons and subduction zones.

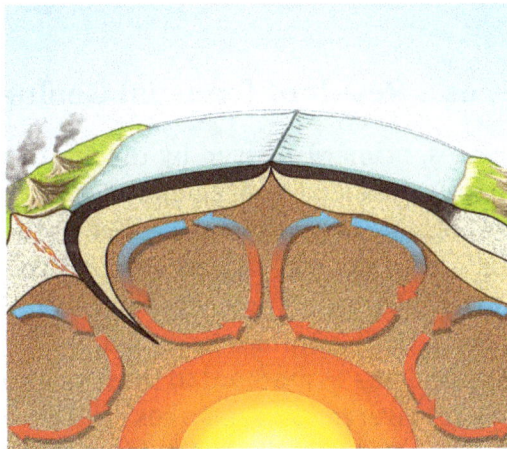

Internal heat engine

## Geological Features as a Result of Internal Geological Process

Internal geological processes like plate convergence, plate divergence and volcanic eruptions tend to lift and modify Earth's relief. The plate convergence/collision has resulted in mountain chains; the mighty Himalaya was created when the Indian and the Eurasian continental plates collided some 40 million years ago (Figure). All along the western coast of South America the plate convergence in the form of collision of oceanic and continental plates resulted in the formation of Andes Mountains. Volcanic eruptions have led to the formation of Deccan Plateau during the Cretaceous period when Indian plate passed over Reunion Plume during its northward journey. The divergence of lithospheric plates in the oceanic environment results in the formation of mammoth network of mid ocean ridges (Example: The Mid Atlantic Ridge, East Pacific Rise, 90 East Ridge, etc.). East Pacific Rise and Mid Atlantic Ridge are examples of relief on the ocean floors. Large scale magmatism in form of extensive and regional igneous provinces could be explained by intraplate magmatism in the rift environmentsand mantle plume activities (Example: the East African Rift system). The convergence of plates/subduction is responsible for opening of back arc basins giving rise to opening of seas (Example: The Japan sea). The internal geological processes significantly build the Earth's relief.

## Sources of Heat for External Processes

Solar energy is the most important source of heat for all external geological processes operating on the Earth's surface. It creates variations in temperature and pressure in the atmosphere that generates wind, including the monsoon. These processes also complete the hydrologic cycle wherein the water changes its domain and forms; from the oceans, rivers lakes and even glaciers it gets evaporated and is reintroduced into the Earth system through various forms of precipitation such as rain and snow. The activity by rain, wind, snow, glaciers, rivers and oceans promote geological processes on the Earth's surface.

## Geological Features as a Result of External Geological Processes

External geological processes, primarily driven by the solar energy constantly shape the Earth's surface at variable rates. External agents such as water, ice, wind and also the human beings are responsible for the changing face of the Earth and modify the existing relief. Highlands are eroded by various external geological agents and the eroded materials are transferred/transported to lowlands and thus result in infilling of the basins and depressions on Earth's surface leading to the formation of sedimentary rocks. These activities over millions of years led to the formation of landforms such as deep valleys, deltas, sand bars, river terraces, beaches etc. Different places on the Earth experience disparate rates of external geological processes. The factors which control the formation of landforms are the climate, location (latitude and longitude), rate of weathering, erosion, transportation, deposition, prevalent lithology and relief of a given place. The processes may operate for millions of years before changing landforms. Sedimentary rocks, particularly those enclosing fossils are important in deciphering relative ages of rocks. These sedimentary rocks also preserve the earliest life records and are very crucial in understanding the Earth's palaeo-environments and evolution of diverse life forms.

## Processes

Geomorphically relevant processes generally fall into (1) the production of regolith by weathering and erosion, (2) the transport of that material, and (3) its eventual deposition. Primary surface processes responsible for most topographic features include wind, waves, chemical dissolution, mass wasting, groundwater movement, surface water flow, glacial action, tectonism, and volcanism. Other more exotic geomorphic processes might include periglacial (freeze-thaw) processes, salt-mediated action, marine currents activity, seepage of fluids through the seafloor or extraterrestrial impact.

Gorge cut by the Indus river into bedrock, Nanga Parbat region, Pakistan. This is the deepest river canyon in the world. Nanga Parbat itself, the world's 9th highest mountain, is seen in the background.

## Aeolian Processes

Wind-eroded alcove near Moab, Utah

Aeolian processes pertain to the activity of the winds and more specifically, to the winds' ability to shape the surface of the Earth. Winds may erode, transport, and deposit materials, and are effective agents in regions with sparse vegetation and a large supply of fine, unconsolidated sediments. Although water and mass flow tend to mobilize more material than wind in most environments, aeolian processes are important in arid environments such as deserts.

## Biological Processes

The interaction of living organisms with landforms, or biogeomorphologic processes, can be of many different forms, and is probably of profound importance for the terrestrial geomorphic system as a whole. Biology can influence very many geomorphic processes, ranging from biogeochemical processes controlling chemical weathering, to the influence of mechanical processes like burrowing and tree throw on soil development, to even controlling global erosion rates through modulation of climate through carbon dioxide balance. Terrestrial landscapes in which the role of biology in mediating surface processes can be definitively excluded are extremely rare, but may hold important information for understanding the geomorphology of other planets, such as Mars.

Beaver dams, as this one in Tierra del Fuego, constitute a specific
form of zoogeomorphology, a type of biogeomorphology.

## Fluvial Processes

Rivers and streams are not only conduits of water, but also of sediment. The water, as it
flows over the channel bed, is able to mobilize sediment and transport it downstream,
either as bed load, suspended load or dissolved load. The rate of sediment transport
depends on the availability of sediment itself and on the river's discharge. Rivers are
also capable of eroding into rock and creating new sediment, both from their own beds
and also by coupling to the surrounding hillslopes. In this way, rivers are thought of as
setting the base level for large-scale landscape evolution in nonglacial environments.
Rivers are key links in the connectivity of different landscape elements.

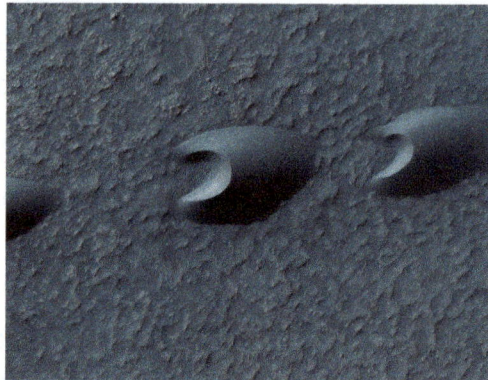

Seif and barchan dunes in the Hellespontus region on the surface of Mars.
Dunes are mobile landforms created by the transport of large volumes of sand by wind.

As rivers flow across the landscape, they generally increase in size, merging with other
rivers. The network of rivers thus formed is a drainage system. These systems take on
four general patterns: dendritic, radial, rectangular, and trellis. Dendritic happens to
be the most common, occurring when the underlying stratum is stable (without faulting). Drainage systems have four primary components: drainage basin, alluvial valley,
delta plain, and receiving basin. Some geomorphic examples of fluvial landforms are
alluvial fans, oxbow lakes, and fluvial terraces.

# Glacial Processes

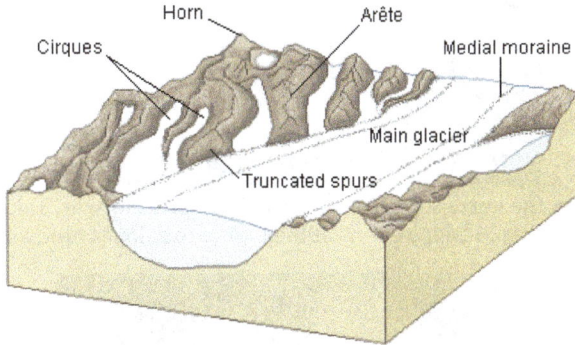

Features of a glacial landscape

Glaciers, while geographically restricted, are effective agents of landscape change. The gradual movement of ice down a valley causes abrasion and plucking of the underlying rock. Abrasion produces fine sediment, termed glacial flour. The debris transported by the glacier, when the glacier recedes, is termed a moraine. Glacial erosion is responsible for U-shaped valleys, as opposed to the V-shaped valleys of fluvial origin.

The way glacial processes interact with other landscape elements, particularly hillslope and fluvial processes, is an important aspect of Plio-Pleistocene landscape evolution and its sedimentary record in many high mountain environments. Environments that have been relatively recently glaciated but are no longer may still show elevated landscape change rates compared to those that have never been glaciated. Nonglacial geomorphic processes which nevertheless have been conditioned by past glaciation are termed paraglacial processes. This concept contrasts with periglacial processes, which are directly driven by formation or melting of ice or frost.

# Hillslope Processes

Soil, regolith, and rock move downslope under the force of gravity via creep, slides, flows, topples, and falls. Such mass wasting occurs on both terrestrial and submarine slopes, and has been observed on Earth, Mars, Venus, Titan and Iapetus.

Ongoing hillslope processes can change the topology of the hillslope surface, which in turn can change the rates of those processes. Hillslopes that steepen up to certain critical thresholds are capable of shedding extremely large volumes of material very quickly, making hillslope processes an extremely important element of landscapes in tectonically active areas.

On the Earth, biological processes such as burrowing or tree throw may play important roles in setting the rates of some hillslope processes.

Talus cones on the north shore of Isfjorden, Svalbard, Norway. Talus cones are accumulations of coarse hillslope debris at the foot of the slopes producing the material.

The Ferguson Slide is an active landslide in the Merced River canyon on California State Highway 140, a primary access road to Yosemite National Park.

## Igneous Processes

Both volcanic (eruptive) and plutonic (intrusive) igneous processes can have important impacts on geomorphology. The action of volcanoes tends to rejuvenize landscapes, covering the old land surface with lava and tephra, releasing pyroclastic material and forcing rivers through new paths. The cones built by eruptions also build substantial new topography, which can be acted upon by other surface processes. Plutonic rocks intruding then solidifying at depth can cause both uplift or subsidence of the surface, depending on whether the new material is denser or less dense than the rock it displaces.

## Tectonic Processes

Tectonic effects on geomorphology can range from scales of millions of years to minutes or less. The effects of tectonics on landscape are heavily dependent on the nature of the underlying bedrock fabric that more or less controls what kind of local morphology tectonics can shape. Earthquakes can, in terms of minutes, submerge large areas of land creating new wetlands. Isostatic rebound can account for significant changes over hundreds to thousands of years, and allows erosion of a mountain belt to promote further erosion as mass is removed from the chain and the belt uplifts. Long-term plate tectonic dynamics give rise to orogenic belts, large mountain chains with typical lifetimes of many tens of millions of years, which form focal points for high rates of fluvial and hillslope processes and thus long-term sediment production.

Features of deeper mantle dynamics such as plumes and delamination of the lower lithosphere have also been hypothesised to play important roles in the long term (> million year), large scale (thousands of km) evolution of the Earth's topography. Both can promote surface uplift through isostasy as hotter, less dense, mantle rocks displace cooler, denser, mantle rocks at depth in the Earth.

## Marine Processes

Marine processes are those associated with the action of waves, marine currents and seepage of fluids through the seafloor. Mass wasting and submarine landsliding are also important processes for some aspects of marine geomorphology. Because ocean basins are the ultimate sinks for a large fraction of terrestrial sediments, depositional processes and their related forms (e.g., sediment fans, deltas) are particularly important as elements of marine geomorphology.

## References

- Dalrymple, G.Brent (2004). Ancient Earth, Ancient Skies: The Age of Earth and Its Cosmic Surroundings. Stanford University Press. p. 52. ISBN 0804749337

- Wilson, J. Tuzo (1965). "A new class of faults and their bearing on continental drift". Nature. 207 (4995): 343–47. Bibcode:1965Natur.207..343W. doi:10.1038/207343a0

- Herbert, Sandra. Charles Darwin as a prospective geological author, British Journal for the History of Science 24. 1991. pp. 159–92

- Second J A (1986) Controversy in Victorian Geology: The Cambrian-Silurian Dispute Princeton University Press, 301 pp. ISBN 0-691-02441-3

- Wilson, J.T> (1963). "Hypothesis on the Earth's behaviour". Nature. 198 (4884): 849–65. Bibcode:1963Natur.198..849H. doi:10.1038/198849a0

- "International Chronostratigraphic Chart v.2013/01" (PDF). International Commission on Stratigraphy. January 2013. Retrieved April 6, 2013

- Rogers, J. J. W. (1996). "A history of continents in the past three billion years". Journal of Geology. 104: 91–107. Bibcode:1996JG....104...91R. doi:10.1086/629803. JSTOR 30068065

- Rudwick, M. J. S. (1985). The Meaning of Fossils: Episodes in the History of Palaeontology. University of Chicago Press. p. 24. ISBN 0-226-73103-0

- International Commission on Stratigraphy. "Chronostratigraphic Units". International Stratigraphic Guide. Archived from the original on 9 December 2009. Retrieved 14 December 2009

- Cox, Simon J. D.; Richard, Stephen M. "A geologic timescale ontology and service". Earth Science Informatics. 8: 5–19. doi:10.1007/s12145-014-0170-6

# 2

# Petrology: A Comprehensive Study

As a sub-discipline of geology, petrology refers to the study of the formation of rocks. Petrology is a vast subject that branches out into significant sub-disciplines, such as igneous petrology, sedimentary petrology and metamorphic petrology, which have been thoroughly discussed in this chapter. It also covers in extensive details the fundamental rock types and the concept of rock cycle.

Petrology is the branch of geology that studies rocks and the conditions under which they are formed. It deals with the origin, occurrence, structure, composition and history of rocks. The term 'petrology' is derived from the two Greek words *pétros* and *logos* meaning "rock" and 'discourse or explanation' respectively.

Rock is a coherent, naturally occurring solid, consisting of an aggregate of minerals. Let us analyse the different aspects of this definition.

- Coherent: Minerals are held in a rock together and can be separated. However, a pile of unattached mineral grains, e.g. loose sand does not constitute a rock.

- Naturally occurring: Materials like brick do not qualify to be a rock. Rock has to be naturally occurring.

- Aggregate of minerals: You have read that rocks consist of aggregate of mineral grains grown or stuck together.

Let us understand the definition of rock with the help of an example. Look carefully at granite with a hand lens (Figure). Granite rock comprises primarily of quartz, K-feldspar, plagioclase feldspar, biotite and mica or hornblende. There are some translucent portions in granite which are constituted of mineral quartz. Flesh pink mineral with tabular habit and pearly luster is K-feldspar (potash feldspar). White mineral with tabular habit and striations, shows properties of plagioclase feldspar. While the brown mineral occurring in thin sheets or flakes and showing pearly luster is mineral biotite mica. You may also find dark coloured mineral with stubby crystals which is probably hornblende. Other minerals like zircon and sphene can occur in granite, but they are not as common as those mentioned above.

Each of the constituent mineral retains its properties in the aggregate of minerals that comprise a rock. A few rocks are composed of non-mineral matters. Coal is considered as rock as it often occurs in layered structure although it consists of organic material. Obsidian and pumice are considered as volcanic rocks even though they are made of glassy material.

Constituent minerals in a granite rock

## Methodology

Petrology utilizes the fields of mineralogy, petrography, optical mineralogy, and chemical analysis to describe the composition and texture of rocks. Petrologists also include the principles of geochemistry and geophysics through the study of geochemical trends and cycles and the use of thermodynamic data and experiments in order to better understand the origins of rocks.

## Branches

There are three branches of petrology, corresponding to the three types of rocks: igneous, metamorphic, and sedimentary, and another dealing with experimental techniques:

- Igneous petrology focuses on the composition and texture of igneous rocks (rocks such as granite or basalt which have crystallized from molten rock or magma). Igneous rocks include volcanic and plutonic rocks.

- Sedimentary petrology focuses on the composition and texture of sedimentary rocks (rocks such as sandstone, shale, or limestone which consist of pieces or particles derived from other rocks or biological or chemical deposits, and are usually bound together in a matrix of finer material).

- Metamorphic petrology focuses on the composition and texture of metamorphic rocks (rocks such as slate, marble, gneiss, or schist which started out as sedimentary or igneous rocks but which have undergone chemical, mineralogical or textural changes due to extremes of pressure, temperature or both).

- Experimental petrology employs high-pressure, high-temperature apparatus to investigate the geochemistry and phase relations of natural or synthetic materials at elevated pressures and temperatures. Experiments are particularly useful for investigating rocks of the lower crust and upper mantle that rarely survive the journey to the surface in pristine condition. They are also one of the prime sources of information about completely inaccessible rocks such as those in the Earth's lower mantle and in the mantles of the other terrestrial planets and the Moon. The work of experimental petrologists has laid a foundation on which modern understanding of igneous and metamorphic processes has been built.

## Igneous Petrology

Igneous petrology is the study of igneous rocks—those that are formed from magma. As a branch of geology, igneous petrology is closely related to volcanology, tectonophysics, and petrology in general. The modern study of igneous rocks utilizes a number of techniques, some of them developed in the fields of chemistry, physics, or other earth sciences. Petrography, crystallography, and isotopic studies are common methods used in igneous petrology.

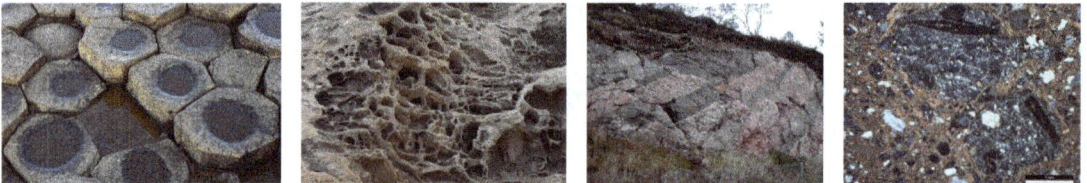

## Methods

### Determination of Chemical Composition

The composition of igneous rocks and minerals can be determined via a variety of methods of varying ease, cost, and complexity. The simplest method is observation of hand samples with the naked eye and/or with a hand lens. This can be used to gauge the general mineralogical composition of the rock, which gives an insight into the composition. A more precise but still relatively inexpensive way to identify minerals (and thereby the bulk chemical composition of the rock) with a petrographic microscope. These microscopes have polarizing plates, filters, and a conoscopic lens that allow the user to measure a large number of crystallographic properties. Another method for determining mineralogy is to use X-ray diffraction, in which a powdered sample is bombarded by X-rays, and the resultant spectrum of crystallographic orientations is compared to a set of standards. One of the most precise ways of determining chemical composition is by the use of an electron microprobe, in which tiny spots of materials are sampled. Electron microprobe analyses can detect both bulk composition and trace element composition.

## Dating Methods

The dating of igneous rocks determines when magma solidified into rock. Radiogenic isotopes are frequently used to determine the age of igneous rocks.

## Potassium–Argon Dating

In this dating method the amount of $^{40}Ar$ trapped in a rock is compared to the amount of $^{40}K$ in the rock to calculate the amount of time $^{40}K$ must have been decaying in the solid rock to produce all $^{40}Ar$ that would have otherwise not have been present there.

## Rubidium–Strontium Dating

The rubidium–strontium dating is based on the natural decay of $^{87}Rb$ to $^{87}Sr$ and the different behaviour of these elements during fractional crystallization of magma. Both Sr and Rb are found in most magmas; however, as fractional crystallization occurs, Sr will tend to be concentrated in plagioclase crystals while Rb will remain in the melt for a longer time. $^{87}Rb$ decays in magma and elsewhere so that every $1.42 \times 10^{11}$ years half of the amount has been converted into $^{87}Sr$. Knowing the decay constant and the amount of $^{87}Rb$ and $^{87}Sr$ in a rock it is possible to calculate the time that the $^{87}Rb$ must have needed before the rock reached closure temperature to produce all $^{87}Sr$, yet considering that there was an initial $^{87}Sr$ amount not produced by $^{87}Rb$ in the magmatic body. Initial values of $^{87}Sr$, when the magma started fractional crystallization, might be estimated by knowing the amounts of $^{87}Rb$ and $^{87}Sr$ of two igneous rocks produced at different times by the same magmatic body.

## Other Methods

Stratigraphic principles may be useful to determine the relative age of volcanic rocks. Tephrochronology is the most common application of stratigraphic dating on volcanic rocks.

## Thermobarometry Methods

In petrology the mineral clinopyroxene is used for temperature and pressure calculations of the magma that produced igneous rock containing this mineral. Clinopyroxene thermobarometry is one of several geothermobarometers. Two things make this method especially useful: first, clinopyroxene is a common phenocryst in igneous rocks easy to identify; and secondly, the crystallization of the jadeite component of clinopyroxene implies a growth in molar volume being thus a good indicator of pressure.

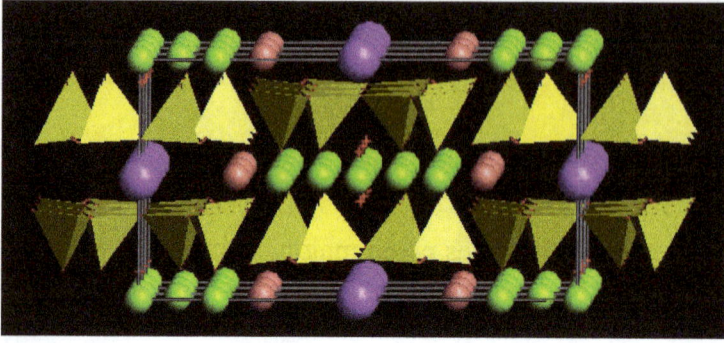

## Sedimentary Petrology

The field of sedimentary petrology is concerned with the description and classification of sedimentary rocks, interpretation of the processes of transportation and deposition of the sedimentary materials forming the rocks, the environment that prevailed at the time the sediments were deposited, and the alteration (compaction, cementation, and chemical and mineralogical modification) of the sediments after deposition.

A banded-iron formation (BIF) rock recovered from the Temagami greenstone belt in Ontario, Canada, and dated to 2.7 billion years ago. Dark layers of iron oxide are intercalated with red chert.

There are two main branches of sedimentary petrology. One branch deals with carbonate rocks, namely limestones and dolomites, composed principally of calcium carbonate (calcite) and calcium magnesium carbonate (dolomite). Much of the complexity in classifying carbonate rocks stems partly from the fact that many limestones and dolomites have been formed, directly or indirectly, through the influence of organisms, including bacteria, lime-secreting algae, various shelled organisms (e.g., mollusks and brachiopods), and by corals. In limestones and dolomites that were deposited under marine conditions, commonly in shallow warm seas, much of the material initially forming the rock consists of skeletons of lime-secreting organisms. In many examples, this skeletal material is preserved as fossils. Some of the major problems of carbonate petrology concern the physical and biological conditions of the environments in which carbonate material has been deposited, including water depth, temperature, degree of illumination by sunlight, motion by waves and currents, and the salinity and other chemical aspects of the water in which deposition occurred.

The other principal branch of sedimentary petrology is concerned with the sediments and sedimentary rocks that are essentially noncalcareous. These include sands and sandstones, clays and claystones, siltstones, conglomerates, glacial till, and varieties of sandstones, siltstones, and conglomerates (e.g., the graywacke-type sandstones and siltstones). These rocks are broadly known as clastic rocks because they consist of distinct particles or clasts. Clastic petrology is concerned with classification, particularly with respect to the mineral composition of fragments or particles, as well as the shapes of particles (angular versus rounded), and the degree of homogeneity of particle sizes. Other main concerns of clastic petrology are the mode of transportation of sedimentary materials, including the transportation of clay, silt, and fine sand by wind; and the transportation of these and coarser materials through suspension in water, through traction by waves and currents in rivers, lakes, and seas, and sediment transport by ice.

Sedimentary petrology also is concerned with the small-scale structural features of sediments and sedimentary rocks. Features that can be conveniently seen in a specimen held in the hand are within the domain of sedimentary petrology. These features include the geometrical attitude of mineral grains with respect to each other, small-scale cross stratification, the shapes and interconnections of pore spaces, and the presence of fractures and veinlets.

Instruments and methods used by sedimentary petrologists include the petrographic microscope for description and classification, X-ray mineralogy for defining fabrics and small-scale structures, physical model flume experiments for studying the effects of flow as an agent of transport and the development of sedimentary structures, and mass spectrometry for calculating stable isotopes and the temperatures of deposition, cementation, and diagenesis. Wet-suit diving permits direct observation of current processes on coral reefs, and manned submersibles enable observation at depth on the ocean floor and in mid-oceanic ridges.

The plate-tectonic theory has given rise to much interest in the relationships between sedimentation and tectonics, particularly in modern plate-tectonic environments—e.g., spreading-related settings (intracontinental rifts, early stages of intercontinental rifting such as the Red Sea, and late stages of intercontinental rifting such as the margins of the present Atlantic Ocean), mid-oceanic settings (ridges and transform faults), subduction-related settings (volcanic arcs, fore-arcs, back-arcs, and trenches), and continental collision-related settings (the Alpine-Himalayan belt and late orogenic basins with molasse [i.e., thick association of clastic sedimentary rocks consisting chiefly of sandstones and shales]). Today many subdisciplines of sedimentary petrology are concerned with the detailed investigation of the various sedimentary processes that occur within these plate-tectonic environments.

## Metamorphic Petrology

Metamorphism means change in form. In geology the term is used to refer to a solid-state recrystallization of earlier igneous, sedimentary, or metamorphic rocks. There are two main types of metamorphism: (1) contact metamorphism, in which changes induced largely by increase in temperature are localized at the contacts of igneous intrusions; and (2) regional metamorphism, in which increased pressure and temperature have caused recrystallization over extensive regions in mountain belts. Other types of metamorphism include local effects caused by deformation in fault zones, burning oil shales, and thrusted ophiolite complexes; extensive recrystallization caused by high heat flow in mid-ocean ridges; and shock metamorphism induced by high-pressure impacts of meteorites in craters on the Earth and Moon.

Metamorphic petrology is concerned with field relations and local tectonic environments; the description and classification of metamorphic rocks in terms of their texture and chemistry, which provides information on the nature of the premetamorphic material; the study of minerals and their chemistry (the mineral assemblages and their possible reactions), which yields data on the temperatures and pressures at which the rocks recrystallized; and the study of fabrics and the relations of mineral growth to deformation stages and major structures, which provides information about the tectonic conditions under which regional metamorphic rocks formed.

A supplement to metamorphism is metasomatism: the introduction and expulsion of fluids and elements through rocks during recrystallization. When new crust is formed and metamorphosed at a mid-oceanic ridge, seawater penetrates into the crust for a few kilometres and carries much sodium with it. During formation of a contact metamorphic aureole around a granitic intrusion, hydrothermal fluids carrying elements such as iron, boron, and fluorine pass from the granite into the wall rocks. When the continental crust is thickened, its lower part may suffer dehydration and form granulites. The expelled fluids, carrying such heat-producing elements as rubidium, uranium, and thorium migrate upward into the upper crust. Much petrologic research is concerned with determining the amount and composition of fluids that have passed through rocks during these metamorphic processes.

The basic instrument used by the metamorphic petrologist is the petrographic microscope, which allows detailed study and definition of mineral types, assemblages, and reactions. If a heating/freezing stage is attached to the microscope, the temperature of formation and composition of fluid inclusions within minerals can be calculated. These inclusions are remnants of the fluids that passed through the rocks during the final stages of their recrystallization. The electron microprobe is widely used for analyzing the composition of the component minerals. The petrologist can combine the mineral chemistry with data from experimental studies and thermodynamics to calculate the pressures and temperatures at which the rocks recrystallized. By obtaining information on the isotopic age of successive metamorphic events with a mass spectrometer, pressure–temperature–time curves can be worked out. These curves chart the movement of the rocks over time as they were brought to the surface from deep within the continental crust; this technique is important for understanding metamorphic processes. Some continental metamorphic rocks that contain diamonds and coesites (ultrahigh pressure minerals) have been carried down subduction zones to a depth of at least 100 kilometres (60 miles), brought up, and often exposed at the present surface within resistant eclogites of collisional orogenic belts—such as the Swiss Alps, the Himalayas, the Kokchetav metamorphic terrane in Kazakhstan, and the Variscan belt in Germany. These examples demonstrate that metamorphic petrology plays a key role in unraveling tectonic processes in mountain belts that have passed through the plate-tectonic cycle of events.

## Rock

Rocks occur on the Earth's surface either as broken chunks (pebbles, cobbles or boulders) that have moved down along slope or by being transported in ice, water or wind or as bedrock that is still attached to the Earth's crust. Geologists refer to an exposure of bedrock as an outcrop that may appear in the field as ridge or cliff, along road/railway cuttings or *nala* or stream cutting.

(a)  (b)  (c)

Field photographs of outcrops along (a) ridge, (b) along road, and (c) river valley

Rock is a naturally occurring solid aggregate of minerals. They provide a historical record of geologic events which give insight into interactions among components (crust and mantle) and spheres (lithosphere, atmosphere, hydrosphere and biosphere) of the Earth System. Rocks occur in a range of colours and textures.

## Classification

Rocks are composed of grains of minerals, which are homogeneous solids formed from a chemical compound arranged in an orderly manner. The aggregate minerals forming the rock are held together by chemical bonds. The types and abundance of minerals in a rock are determined by the manner in which it was formed.

Rock outcrop along a mountain creek near Orosí, Costa Rica.

Many rocks contain silica (SiO$_2$); a compound of silicon and oxygen that forms 74.3% of the Earth's crust. This material forms crystals with other compounds in the rock. The proportion of silica in rocks and minerals is a major factor in determining their names and properties.

Rocks are classified according to characteristics such as mineral and chemical composition, permeability, texture of the constituent particles, and particle size. These physical properties are the result of the processes that formed the rocks. Over the course of time, rocks can transform from one type into another, as described by a geological model called the rock cycle. This transformation produces three general classes of rock: igneous, sedimentary, and metamorphic.

Those three classes are subdivided into many groups. There are, however, no hard-and-fast boundaries between allied rocks. By increase or decrease in the proportions of their minerals, they pass through gradations from one to the other; the distinctive structures of one kind of rock may thus be traced gradually merging into those of another. Hence the definitions adopted in rock names simply correspond to selected points in a continuously graduated series.

## Igneous Rock

Sample of igneous gabbro

Igneous rock (derived from the Latin word *igneus,* meaning *of fire,* from *ignis* meaning *fire)* is formed through the cooling and solidification of magma or lava. This magma may be derived from partial melts of pre-existing rocks in either a planet's mantle or crust. Typically, the melting of rocks is caused by one or more of three processes: an increase in temperature, a decrease in pressure, or a change in composition.

Igneous rocks are divided into two main categories:

- Plutonic or intrusive rocks result when magma cools and crystallizes slowly within the Earth's crust. A common example of this type is granite.

- Volcanic or extrusive rocks result from magma reaching the surface either as lava or *fragmental ejecta*, forming minerals such as pumice or basalt.

The chemical abundance and the rate of cooling of magma typically forms a sequence known as Bowen's reaction series. Most major igneous rocks are found along this scale.

About 64.7% of the Earth's crust by volume consists of igneous rocks, making it the most plentiful category. Of these, 66% are basalts and gabbros, 16% are granite, and 17% granodiorites and diorites. Only 0.6% are syenites and 0.3% peridotites and dunites. The oceanic crust is 99% basalt, which is an igneous rock of mafic composition. Granites and similar rocks, known as meta-granitoids, form much of the continental crust. Over 700 types of igneous rocks have been described, most of them having formed beneath the surface of Earth's crust. These have diverse properties, depending on their composition and the temperature and pressure conditions in which they were formed.

## Sedimentary Rock

Sedimentary sandstone with iron oxide bands

Sedimentary rocks are formed at the earth's surface by the accumulation and cementation of fragments of earlier rocks, minerals, and organisms or as chemical precipitates and organic growths in water (sedimentation). This process causes clastic sediments (pieces of rock) or organic particles (detritus) to settle and accumulate, or for minerals to chemically precipitate (evaporite) from a solution. The particulate matter then undergoes compaction and cementation at moderate temperatures and pressures (diagenesis).

Before being deposited, sediments are formed by weathering of earlier rocks by erosion in a source area and then transported to the place of deposition by water, wind, ice, mass movement or glaciers (agents of denudation). Mud rocks comprise 65% (mudstone, shale and siltstone); sandstones 20 to 25% and carbonate rocks 10 to 15% (limestone and dolostone). About 7.9% of the crust by volume is composed of sedimentary rocks, with 82% of those being shales, while the remainder consists of limestone (6%), sandstone and arkoses (12%). Sedimentary rocks often contain fossils. Sedimentary rocks form under the influence of gravity and typically are deposited in horizontal or near horizontal layers or strata and may be referred to as stratified rocks. A small fraction of sedimentary rocks deposited on steep slopes will show cross bedding where one layer stops abruptly along an interface where another layer eroded the first as it was laid atop the first.

## Metamorphic Rock

Metamorphic banded gneiss

Metamorphic rocks are formed by subjecting any rock type—sedimentary rock, igneous rock or another older metamorphic rock—to different temperature and pressure conditions than those in which the original rock was formed. This process is called metamorphism, meaning to "change in form". The result is a profound change in physical properties and chemistry of the stone. The original rock, known as the protolith, transforms into other mineral types or other forms of the same minerals, by recrystallization. The temperatures and pressures required for this process are always higher than those found at the Earth's surface: temperatures greater than 150 to 200 °C and pressures of 1500 bars. Metamorphic rocks compose 27.4% of the crust by volume.

The three major classes of metamorphic rock are based upon the formation mechanism. An intrusion of magma that heats the surrounding rock causes contact metamorphism—a temperature-dominated transformation. Pressure metamorphism occurs when sediments are buried deep under the ground; pressure is dominant, and temperature plays a smaller role. This is termed burial metamorphism, and it can result in rocks such as jade. Where both heat and pressure play a role, the mechanism is termed regional metamorphism. This is typically found in mountain-building regions.

Depending on the structure, metamorphic rocks are divided into two general categories. Those that possess a texture are referred to as foliated; the remainders are termed non-foliated. The name of the rock is then determined based on the types of minerals present. Schists are foliated rocks that are primarily composed of lamellar minerals such as micas. A gneiss has visible bands of differing lightness, with a common example being the granite gneiss. Other varieties of foliated rock include slates, phyllites, and mylonite. Familiar examples of non-foliated metamorphic rocks include marble, soapstone, and serpentine. This branch contains quartzite—a metamorphosed form of sandstone—and hornfels.

# Igneous Rocks

Igneous rocks (word derived from the Latin *ignis,* meaning "fire" or "to ignite") are formed by cooling and crystallisation of hot molten material called magma, which rises up from the mantle inside Earth and cools. This cooling may happen either below or above the surface of the Earth. Thus depending upon whether the cooling and crystallization of igneous rocks took place beneath or above the surface of the Earth. They may be grouped into intrusive or extrusive igneous rocks respectively.

## Geological Significance

Igneous and metamorphic rocks make up 90–95% of the top 16 km of the Earth's crust by volume. Igneous rocks form about 15% of the Earth's current land surface. Most of the Earth's oceanic crust is made of igneous rock.

Igneous rocks are also geologically important because:

- their minerals and global chemistry give information about the composition of the mantle, from which some igneous rocks are extracted, and the temperature and pressure conditions that allowed this extraction, and/or of other pre-existing rock that melted;

- their absolute ages can be obtained from various forms of radiometric dating and thus can be compared to adjacent geological strata, allowing a time sequence of events;

- their features are usually characteristic of a specific tectonic environment, allowing tectonic reconstitutions;

- in some special circumstances they host important mineral deposits (ores): for example, tungsten, tin, and uranium are commonly associated with granites and diorites, whereas ores of chromium and platinum are commonly associated with gabbros.

## Geological Setting

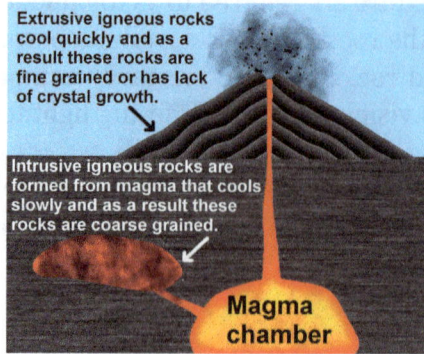

Forming of igneous rock

In terms of modes of occurrence, igneous rocks can be either intrusive (plutonic and hypabyssal) or extrusive (volcanic).

## Intrusive

Close-up of granite (an intrusive igneous rock) exposed in Chennai, India.

Intrusive igneous rocks are formed from magma that cools and solidifies within the crust of a planet, surrounded by pre-existing rock (called country rock); the magma cools slowly and, as a result, these rocks are coarse-grained. The mineral grains in such rocks can generally be identified with the naked eye. Intrusive rocks can also be classified according to the shape and size of the intrusive body and its relation to the other formations into which it intrudes. Typical intrusive formations are batholiths, stocks, laccoliths, sills and dikes. When the magma solidifies within the earth's crust, it cools slowly forming coarse textured rocks, such as granite, gabbro, or diorite.

The central cores of major mountain ranges consist of intrusive igneous rocks, usually granite. When exposed by erosion, these cores (called *batholiths*) may occupy huge areas of the Earth's surface.

Intrusive igneous rocks that form at depth within the crust are termed plutonic (or *abyssal*) rocks and are usually coarse-grained. Intrusive igneous rocks that form near the surface are termed *subvolcanic* or *hypabyssal* rocks and they are usually medium-grained. Hypabyssal rocks are less common than plutonic or volcanic rocks and often form dikes, sills, laccoliths, lopoliths, or phacoliths.

## Extrusive

Extrusive igneous rock is made from lava released by volcanoes

Sample of basalt (an extrusive igneous rock), found in Massachusetts

Extrusive igneous rocks, also known as volcanic rocks, are formed at the crust's surface as a result of the partial melting of rocks within the mantle and crust. Extrusive igneous rocks cool and solidify quicker than intrusive igneous rocks. They are formed by the cooling of molten magma on the earth's surface. The magma, which is brought to the surface through fissures or volcanic eruptions, solidifies at a faster rate. Hence such rocks are smooth, crystalline and fine-grained. Basalt is a common extrusive igneous rock and forms lava flows, lava sheets and lava plateaus. Some kinds of basalt solidify to form long polygonal columns. The Giant's Causeway in Antrim, Northern Ireland is an example.

The molten rock, with or without suspended crystals and gas bubbles, is called magma. It rises because it is less dense than the rock from which it was created. When magma reaches the surface from beneath water or air, it is called lava. Eruptions of volcanoes into air are termed *subaerial*, whereas those occurring underneath the ocean are termed *submarine*. Black smokers and mid-ocean ridge basalt are examples of submarine volcanic activity.

The volume of extrusive rock erupted annually by volcanoes varies with plate tectonic setting. Extrusive rock is produced in the following proportions:

- divergent boundary: 73%

- convergent boundary (subduction zone): 15%

- hotspot: 12%.

Magma that erupts from a volcano behaves according to its viscosity, determined by temperature, composition, crystal content and the amount of silica. High-temperature magma, most of which is basaltic in composition, behaves in a manner similar to thick oil and, as it cools, treacle. Long, thin basalt flows with pahoehoe surfaces are common. Intermediate composition magma, such as andesite, tends to form cinder cones of intermingled ash, tuff and lava, and may have a viscosity similar to thick, cold molasses or even rubber when erupted. Felsic magma, such as rhyolite, is usually erupted at low temperature and is up to 10,000 times as viscous as basalt. Volcanoes with rhyolitic magma commonly erupt explosively, and rhyolitic lava flows are typically of limited extent and have steep margins, because the magma is so viscous.

Felsic and intermediate magmas that erupt often do so violently, with explosions driven by the release of dissolved gases—typically water vapour, but also carbon dioxide. Explosively erupted pyroclastic material is called tephra and includes tuff, agglomerate and ignimbrite. Fine volcanic ash is also erupted and forms ash tuff deposits, which can often cover vast areas.

Because lava usually cools and crystallizes rapidly, it is usually fine-grained. If the cooling has been so rapid as to prevent the formation of even small crystals after extrusion, the resulting rock may be mostly glass (such as the rock obsidian). If the cooling of the lava happened more slowly, the rock would be coarse-grained.

Because the minerals are mostly fine-grained, it is much more difficult to distinguish between the different types of extrusive igneous rocks than between different types of intrusive igneous rocks. Generally, the mineral constituents of fine-grained extrusive igneous rocks can only be determined by examination of thin sections of the rock under a microscope, so only an approximate classification can usually be made in the field.

## Classification

Igneous rocks are classified according to mode of occurrence, texture, mineralogy, chemical composition, and the geometry of the igneous body.

The classification of the many types of different igneous rocks can provide us with important information about the conditions under which they formed. Two important variables used for the classification of igneous rocks are particle size, which largely depends on the cooling history, and the mineral composition of the rock. Feldspars, quartz or feldspathoids, olivines, pyroxenes, amphiboles, and micas are all important minerals in the formation of almost all igneous rocks, and they are basic to the classification of these rocks. All other minerals present are regarded as nonessential in almost all igneous rocks and are called *accessory minerals*. Types of igneous rocks with other essential minerals are very rare, and these rare rocks include those with essential carbonates.

In a simplified classification, igneous rock types are separated on the basis of the type of feldspar present, the presence or absence of quartz, and in rocks with no feldspar or quartz, the type of iron or magnesium minerals present. Rocks containing quartz (silica in composition) are silica-oversaturated. Rocks with feldspathoids are silica-undersaturated, because feldspathoids cannot coexist in a stable association with quartz.

Igneous rocks that have crystals large enough to be seen by the naked eye are called phaneritic; those with crystals too small to be seen are called aphanitic. Generally speaking, phaneritic implies an intrusive origin; aphanitic an extrusive one.

An igneous rock with larger, clearly discernible crystals embedded in a finer-grained matrix is termed porphyry. Porphyritic texture develops when some of the crystals grow to considerable size before the main mass of the magma crystallizes as finer-grained, uniform material.

Igneous rocks are classified on the basis of texture and composition. Texture refers to the size, shape, and arrangement of the mineral grains or crystals of which the rock is composed.

## Texture

Gabbro specimen showing phaneritic texture; Rock Creek Canyon,
eastern Sierra Nevada, California; scale bar is 2.0 cm.

Texture is an important criterion for the naming of volcanic rocks. The texture of volcanic rocks, including the size, shape, orientation, and distribution of mineral grains and the intergrain relationships, will determine whether the rock is termed a tuff, a pyroclastic lava or a simple lava.

However, the texture is only a subordinate part of classifying volcanic rocks, as most often there needs to be chemical information gleaned from rocks with extremely fine-grained groundmass or from airfall tuffs, which may be formed from volcanic ash.

Textural criteria are less critical in classifying intrusive rocks where the majority of minerals will be visible to the naked eye or at least using a hand lens, magnifying glass or microscope. Plutonic rocks also tend to be less texturally varied and less prone to gaining structural fabrics. Textural terms can be used to differentiate different intrusive phases of large plutons, for instance porphyritic margins to large intrusive bodies, porphyry stocks and subvolcanic dikes (apophyses). Mineralogical classification is most often used to classify plutonic rocks. Chemical classifications are preferred to classify volcanic rocks, with phenocryst species used as a prefix, e.g. "olivine-bearing picrite" or "orthoclase-phyric rhyolite".

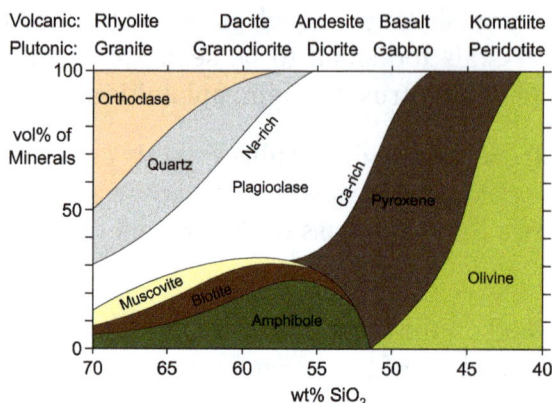

Basic classification scheme for igneous rocks on their mineralogy. If the approximate volume fractions of minerals in the rock are known, the rock name and silica content can be read off the diagram. This is not an exact method, because the classification of igneous rocks also depends on other components than silica, yet in most cases it is a good first guess.

## Chemical Classification and Petrology

Igneous rocks can be classified according to chemical or mineralogical parameters.

Chemical: total alkali-silica content (TAS diagram) for volcanic rock classification used when modal or mineralogic data is unavailable:

- *felsic* igneous rocks containing a high silica content, greater than 63% $SiO_2$ (examples granite and rhyolite),

- *intermediate* igneous rocks containing between 52–63% $SiO_2$ (example andesite and dacite),

- *mafic* igneous rocks have low silica 45–52% and typically high iron – magnesium content (example gabbro and basalt),

- *ultramafic rock* igneous rocks with less than 45% silica (examples picrite, komatiite and peridotite),

- *alkalic* igneous rocks with 5–15% alkali ($K_2O + Na_2O$) content or with a molar ratio of alkali to silica greater than 1:6 (examples phonolite and trachyte).

Chemical classification also extends to differentiating rocks that are chemically similar according to the TAS diagram, for instance:

- Ultrapotassic – rocks containing molar $K_2O/Na_2O$ >3.

- Peralkaline – rocks containing molar $(K_2O + Na_2O)/Al_2O_3 > 1$.

- Peraluminous – rocks containing molar $(K_2O + Na_2O)/Al_2O_3 < 1$.

An idealized mineralogy (the normative mineralogy) can be calculated from the chemical composition, and the calculation is useful for rocks too fine-grained or too altered for identification of minerals that crystallized from the melt. For instance, normative quartz classifies a rock as silica-oversaturated; an example is rhyolite. In an older terminology, silica oversaturated rocks were called *silicic* or *acidic* where the $SiO_2$ was greater than 66% and the family term *quartzolite* was applied to the most silicic. A normative feldspathoid classifies a rock as silica-undersaturated; an example is nephelinite.

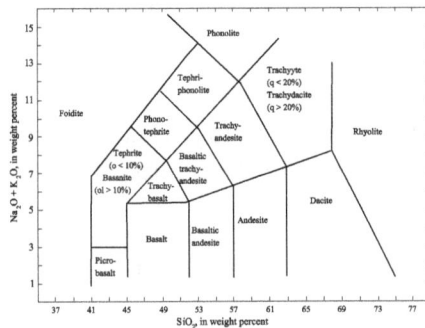

Total alkali versus silica classification scheme (TAS) as proposed in Le Maitre's 2002
Igneous Rocks - A classification and glossary of terms

## History of Classification

In 1902, a group of American petrographers proposed that all existing classifications of igneous rocks should be discarded and replaced by a "quantitative" classification based on chemical analysis. They showed how vague, and often unscientific, much of the existing terminology was and argued that as the chemical composition of an igneous rock was its most fundamental characteristic, it should be elevated to prime position.

Geological occurrence, structure, mineralogical constitution—the hitherto accepted criteria for the discrimination of rock species—were relegated to the background. The completed rock analysis is first to be interpreted in terms of the rock-forming minerals which might be expected to be formed when the magma crystallizes, e.g., quartz feldspars, olivine, akermannite, Feldspathoids, magnetite, corundum, and so on, and the rocks are divided into groups strictly according to the relative proportion of these minerals to one another.

## Mineralogical Classification

For volcanic rocks, mineralogy is important in classifying and naming lavas. The most important criterion is the phenocryst species, followed by the groundmass mineralogy. Often, where the groundmass is aphanitic, chemical classification must be used to properly identify a volcanic rock.

## Mineralogic Contents – Felsic Versus Mafic

- *felsic* rock, highest content of silicon, with predominance of quartz, alkali feldspar and/or feldspathoids: *the felsic minerals*; these rocks (e.g., granite, rhyolite) are usually light coloured, and have low density.

- *mafic* rock, lesser content of silicon relative to felsic rocks, with predominance of mafic minerals pyroxenes, olivines and calcic plagioclase; these rocks (example, basalt, gabbro) are usually dark coloured, and have a higher density than felsic rocks.

- *ultramafic* rock, lowest content of silicon, with more than 90% of mafic minerals (e.g., dunite).

For intrusive, plutonic and usually phaneritic igneous rocks (where all minerals are visible at least via microscope), the mineralogy is used to classify the rock. This usually occurs on ternary diagrams, where the relative proportions of three minerals are used to classify the rock.

The following table is a simple subdivision of igneous rocks according to both their composition and mode of occurrence.

| Mode of occurrence | Composition | | | |
| --- | --- | --- | --- | --- |
| | Felsic | Intermediate | Mafic | Ultramafic |
| Intrusive | Granite | Diorite | Gabbro | Peridotite |
| Extrusive | Rhyolite | Andesite | Basalt | Komatiite |

## Example of Classification

Granite is an igneous intrusive rock (crystallized at depth), with felsic composition (rich in silica and predominately quartz plus potassium-rich feldspar plus sodium-rich plagioclase) and phaneritic, subeuhedral texture (minerals are visible to the unaided eye and commonly some of them retain original crystallographic shapes).

## Magma Origination

The Earth's crust averages about 35 kilometers thick under the continents, but averages only some 7–10 kilometers beneath the oceans. The continental crust is composed primarily of sedimentary rocks resting on a crystalline *basement* formed of a great variety of metamorphic and igneous rocks, including granulite and granite. Oceanic crust is composed primarily of basalt and gabbro. Both continental and oceanic crust rest on peridotite of the mantle.

Rocks may melt in response to a decrease in pressure, to a change in composition (such as an addition of water), to an increase in temperature, or to a combination of these processes.

Other mechanisms, such as melting from a meteorite impact, are less important today, but impacts during the accretion of the Earth led to extensive melting, and the outer several hundred kilometers of our early Earth was probably an ocean of magma. Impacts of large meteorites in the last few hundred million years have been proposed as one mechanism responsible for the extensive basalt magmatism of several large igneous provinces.

## Decompression

Decompression melting occurs because of a decrease in pressure.

The solidus temperatures of most rocks (the temperatures below which they are completely solid) increase with increasing pressure in the absence of water. Peridotite at depth in the Earth's mantle may be hotter than its solidus temperature at some shallower level. If such rock rises during the convection of solid mantle, it will cool slightly as it expands in an adiabatic process, but the cooling is only about 0.3 °C per kilometer. Experimental studies of appropriate peridotite samples document that the solidus temperatures increase by 3 °C to 4 °C per kilometer. If the rock rises far enough, it will begin to melt. Melt droplets can coalesce into larger volumes and be intruded upwards. This process of melting from the upward movement of solid mantle is critical in the evolution of the Earth.

Decompression melting creates the ocean crust at mid-ocean ridges. It also causes volcanism in intraplate regions, such as Europe, Africa and the Pacific sea floor. There, it is variously attributed either to the rise of mantle plumes (the "Plume hypothesis") or to intraplate extension (the "Plate hypothesis").

## Effects of Water and Carbon Dioxide

The change of rock composition most responsible for the creation of magma is the addition of water. Water lowers the solidus temperature of rocks at a given pressure. For example, at a depth of about 100 kilometers, peridotite begins to melt near 800 °C in the presence of excess water, but near or above about 1,500 °C in the absence of water. Water is driven out of the oceanic lithosphere in subduction zones, and it causes melting in the overlying mantle. Hydrous magmas composed of basalt and andesite are produced directly and indirectly as results of dehydration during the subduction process. Such magmas, and those derived from them, build up island arcs such as those in the Pacific Ring of Fire. These magmas form rocks of the calc-alkaline series, an important part of the continental crust.

The addition of carbon dioxide is relatively a much less important cause of magma formation than the addition of water, but genesis of some silica-undersaturated magmas has been attributed to the dominance of carbon dioxide over water in their mantle source regions. In the presence of carbon dioxide, experiments document that the peridotite solidus temperature decreases by about 200 °C in a narrow pressure interval at pressures corresponding to a depth of about 70 km. At greater depths, carbon dioxide can have more effect: at depths to about 200 km, the temperatures of initial melting of a carbonated peridotite composition were determined to be 450 °C to 600 °C lower than for the same composition with no carbon dioxide. Magmas of rock types such as nephelinite, carbonatite, and kimberlite are among those that may be generated following an influx of carbon dioxide into mantle at depths greater than about 70 km.

## Temperature Increase

Increase in temperature is the most typical mechanism for formation of magma within continental crust. Such temperature increases can occur because of the upward intrusion of magma from the mantle. Temperatures can also exceed the solidus of a crustal rock in continental crust thickened by compression at a plate boundary. The plate boundary between the Indian and Asian continental masses provides a well-studied example, as the Tibetan Plateau just north of the boundary has crust about 80 kilometers thick, roughly twice the thickness of normal continental crust. Studies of electrical resistivity deduced from magnetotelluric data have detected a layer that appears to contain silicate melt and that stretches for at least 1,000 kilometers within the middle crust along the southern margin of the Tibetan Plateau. Granite and rhyolite are types of igneous rock commonly interpreted as products of the melting of continental crust because of increases in temperature. Temperature increases also may contribute to the melting of lithosphere dragged down in a subduction zone.

## Magma Evolution

Most magmas only entirely melt for small parts of their histories. More typically, they are mixes of melt and crystals, and sometimes also of gas bubbles. Melt, crystals, and bubbles usually have different densities, and so they can separate as magmas evolve.

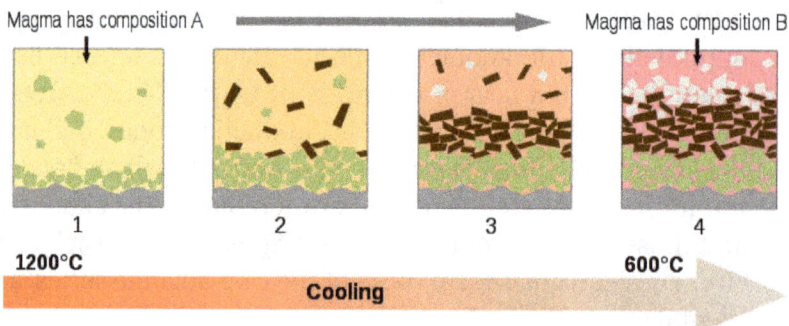

Schematic diagrams showing the principles behind fractional crystallisation in a magma. While cooling, the magma evolves in composition because different minerals crystallize from the melt. 1: olivine crystallizes; 2: olivine and pyroxene crystallize; 3: pyroxene and plagioclase crystallize; 4: plagioclase crystallizes. At the bottom of the magma reservoir, a cumulate rock forms.

As magma cools, minerals typically crystallize from the melt at different tempera-tures (fractional crystallization). As minerals crystallize, the composition of the resid-ual melt typically changes. If crystals separate from the melt, then the residual melt will differ in composition from the parent magma. For instance, a magma of gabbro-ic composition can produce a residual melt of granitic composition if early formed crystals are separated from the magma. Gabbro may have a liquidus temperature near 1,200 °C, and the derivative granite-composition melt may have a liquidus tem-perature as low as about 700 °C. Incompatible elements are concentrated in the last residues of magma during fractional crystallization and in the first melts produced during partial melting: either process can form the magma that crystallizes to peg-matite, a rock type commonly enriched in incompatible elements. Bowen's reaction series is important for understanding the idealised sequence of fractional crystallisa-tion of a magma.

Magma composition can be determined by processes other than partial melting and fractional crystallization. For instance, magmas commonly interact with rocks they in-trude, both by melting those rocks and by reacting with them. Magmas of different compositions can mix with one another. In rare cases, melts can separate into two im-miscible melts of contrasting compositions.

There are relatively few minerals that are important in the formation of common ig-neous rocks, because the magma from which the minerals crystallize is rich in only certain elements: silicon, oxygen, aluminium, sodium, potassium, calcium, iron, and magnesium. These are the elements that combine to form the silicate minerals, which account for over ninety percent of all igneous rocks. The chemistry of igneous rocks is expressed differently for major and minor elements and for trace elements. Contents of major and minor elements are conventionally expressed as weight per-cent oxides (e.g., 51% $SiO_2$, and 1.50% $TiO_2$). Abundances of trace elements are con-ventionally expressed as parts per million by weight (e.g., 420 ppm Ni, and 5.1 ppm Sm). The term "trace element" is typically used for elements present in most rocks at abundances less than 100 ppm or so, but some trace elements may be present in some rocks at abundances exceeding 1,000 ppm. The diversity of rock compositions has been defined by a huge mass of analytical data—over 230,000 rock analyses can be accessed on the web through a site sponsored by the U. S. National Science Foundation.

Intrusive igneous rocks are formed by the cooling and crystallization of magma at depth. Intrusive also known as plutonic rocks crystallise when magma cools in the magma chamber or intrudes the country rocks or the rock bodies enclosing an intru-sive mass of igneous rock. The magma beneath the surface of the Earth undergoes slow cooling resulting in the formation of large crystals giving rise to coarse grained rocks, recognized by their large, interlocking crystals visible in hand specimen, e.g. granite, granodiorite, gabbro, and diorite.

## Formation

Intrusive rock forms within Earth's crust from the crystallization of magma. Many mountain ranges, such as the Sierra Nevada in California, are formed mostly from large granite (or related rock) intrusions; see Sierra Nevada batholith.

## Related Forms

Intrusions are one of the two ways igneous rock can form; the other is extrusive rock, that is, a volcanic eruption or similar event. Technically speaking, an intrusion is any formation of intrusive igneous rock; rock formed from magma that cools and solidifies within the crust of the planet. In contrast, an *extrusion* consists of extrusive rock; rock formed above the surface of the crust.

## Related Terms

Large bodies of magma that solidify underground before they reach the surface of the crust are called plutons. Plutonic rocks form 7% of the Earth's current land surface.

Coarse-grained intrusive igneous rocks that form at depth within the earth are called *abyssal* while those that form near the surface are called subvolcanic or *hypabyssal*. Intrusive structures are often classified according to whether or not they are parallel to the bedding planes or foliation of the country rock: if the intrusion is parallel the body is *concordant*, otherwise it is *discordant*.

## Intrusive Suite

An *intrusive suite* is a group of plutons related in time and space.

## Variations

Intrusions vary widely, from mountain-range-sized batholiths to thin veinlike fracture fillings of aplite or pegmatite.

## Structural Types

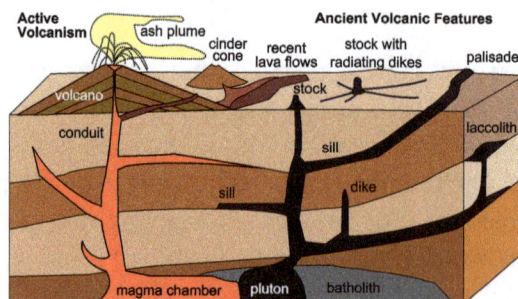

Diagram showing various types of igneous intrusion

A *dike* intrudes into the country rock, Baranof Island, Alaska, United States

Intrusions can be classified according to the shape and size of the intrusive body and its relation to the other formations into which it intrudes:

Batholith: a large irregular discordant intrusion

Chonolith: an irregularly-shaped intrusion with a demonstrable base

Cupola: a dome-shaped projection from the top of a large subterranean intrusion

Dike: a relatively narrow tabular discordant body, often nearly vertical

Laccolith: concordant body with roughly flat base and convex top, usually with a feeder pipe below

Lopolith: concordant body with roughly flat top and a shallow convex base, may have a feeder dike or pipe below

Phacolith: a concordant lens-shaped pluton that typically occupies the crest of an anticline or trough of a syncline

Volcanic pipe or volcanic neck: tubular roughly vertical body that may have been a feeder vent for a volcano

Sill: a relatively thin tabular concordant body intruded along bedding planes

Stock: a smaller irregular discordant intrusive

Boss: a small stock

## Characteristics

A body of intrusive igneous rock which crystallizes from magma cooling underneath the surface of the Earth is called a *pluton*. If the pluton is large, it may be called a *batholith* or a *stock*.

An intrusion (pink Notch Peak monzonite) inter-fingers (partly as a dike) with highly metamorphosed black-and-white-striped host rock (Cambrian carbonate rocks) near Notch Peak, House Range, Utah

Intrusive rocks are characterized by large crystal sizes, and as the individual crystals are visible, the rock is called *phaneritic*. This is as the magma cools underground, and while cooling may be fast or slow, cooling is slower than on the surface, so larger crystals grow. If it runs parallel to rock layers, it is called a *sill*. If an intrusion makes rocks above rise to form a dome, it is called a *laccolith*.

How deep-seated intrusions burst through the overlying strata causes intrusive rock to be recognized: Veins spread out into branches, or branchlike parts result from filled cracks, and the high temperature is evident in how they alter country rock. As heat dissipation is slow, and as the rock is under pressure, crystals form, and no vitreous rapidly chilled matter is present.

The intrusions did not flow while solidifying, hence do not show lines. Contained gases could not escape through the thick strata, thus form cavities, which can often be observed. Because their crystals are of the rough equal size, these rocks are said to be *equigranular*.

There is typically no distinction between a first generation of large well-shaped crystals and a fine-grained ground-mass. The minerals of each have formed in a definite order, and each has had a period of crystallization that may be very distinct or may have coincided with or overlapped the period of formation of some of the other ingredients. Earlier crystals originated at a time when most of the rock was still liquid and are more or less perfect. Later crystals are less regular in shape because they were compelled to occupy the spaces left between the already-formed crystals. The former case is said to be idiomorphic (or *automorphic*); the latter is xenomorphic. There are also many other characteristics that serve to distinguish the members of these two groups. For example, orthoclase is typically feldspar from granite, while its modifications occur in lavas of similar composition. The same distinction holds for nepheline varieties. Leucite is common in lavas but very rare in plutonic rocks. Muscovite is confined to intrusions. These differences show the influence of the physical conditions under which consolidation takes place.

Intrusive rocks formed at greater depths are called plutonic or *abyssal*. Some intrusive rocks solidified in fissures as dikes and intrusive sills at shallow depth and are called *subvolcanic* or *hypabyssal*. They show structures intermediate between those of extrusive and plutonic rocks. They are very commonly porphyritic, vitreous, and sometimes even vesicular. In fact, many of them are petrologically indistinguishable from lavas of similar composition.

Extrusive igneous rocks are also known as volcanic rocks. They are formed when the hot molten material erupts at Earth's surface, spreads out as lava flow and undergoes rapid cooling in the contact with air and water. They may have air cavities or vesicles, indicating that gas has escaped from the site on release of pressure. If the overlying rock has fractures, then the pressure may be released and a sizeable volume of molten rock will extrude to the surface. Extrusive igneous rocks, such as basalt, rhyolite, trachyte are easily recognised by their fine grained or glassy texture.

The igneous rocks based on their mode of occurrence can be classified into three types (Figure) based on their texture:

- Plutonic rocks: The term plutonic is derived from *Pluto* the Roman God of the underworld. These rocks undergo cooling and consolidation beneath the surface of the Earth or with in the magma chamber such as granite, gabbro. Plutonic rocks occur as intrusive bodies like batholith, *e.g.* Mount Abu, Ladakh batholith.

- Volcanic rocks: These rocks undergo cooling and consolidation at the surface of the Earth in contact with air or water such as basalt, rhyolite. They occur as extrusive bodies like lava flow, *e.g.* Deccan basalts, Malani rhyolite, Jodhpur.

- Hypabyssal rocks: These rocks undergo cooling and consolidation at the shallow level/ near the surface of the Earth such as dolerite. They are medium grained and occur often as dykes or sills.

Plutonic, hypabyssal and volcanic equivalents of rock consisting essentially of minerals like clinopyroxene and calcic plagioclase. (a) plutonic rock is gabbro, (b) hypabyssal is dolerite, and (c) volcanic equivalent is basalt.

The most of the minerals present in igneous rocks are silicates, partly because -silica is - abundant in Earth's crust and partly because many silicate minerals melt at the high temperatures and pressures reached in deeper parts of the crust and in the mantle. The silicate minerals most commonly found in igneous rocks include quartz, feldspars, micas, pyroxenes, amphiboles and olivine.

## Sedimentary Rocks

The sedimentary rock is formed at or near the surface of the Earth in one of several ways discussed in section. We can compare the layers of sedimentary rocks to the pages ofbook that record stories of earlier events and environments of our dynamic planet Earth. Sedimentologists are a specific group of geologists who study sedimentary rocks.

### Classification based on Origin

Sedimentary rocks can be subdivided into four groups based on the processes responsible for their formation: clastic sedimentary rocks, biochemical (biogenic) sedimentary rocks, chemical sedimentary rocks, and a fourth category for "other" sedimentary rocks formed by impacts, volcanism, and other minor processes.

### Clastic Sedimentary Rocks

Clastic sedimentary rocks are composed of other rock fragments that were cemented by silicate minerals. Clastic rocks are composed largely of quartz, feldspar, rock (lithic) fragments, clay minerals, and mica; any type of mineral may be present, but they in general represent the minerals that exist locally.

Claystone deposited in Glacial Lake Missoula, Montana, United States. Note the very fine and flat bedding, common for distal lacustrine deposition.

Clastic sedimentary rocks, are subdivided according to the dominant particle size. Most geologists use the Udden-Wentworth grain size scale and divide unconsolidated sediment into three fractions: gravel (>2 mm diameter), sand (1/16 to 2 mm diameter), and mud (clay is <1/256 mm and silt is between 1/16 and 1/256 mm). The classification of clastic sedimentary rocks parallels this scheme; conglomerates and breccias are made mostly of gravel, sandstones are made mostly of sand, and mudrocks are made mostly of the finest material. This tripartite subdivision is mirrored by the broad categories of rudites, arenites, and lutites, respectively, in older literature.

The subdivision of these three broad categories is based on differences in clast shape (conglomerates and breccias), composition (sandstones), grain size or texture (mudrocks).

## Conglomerates and Breccias

Conglomerates are dominantly composed of rounded gravel, while breccias are composed of dominantly angular gravel.

## Sandstones

Sedimentary rock with sandstone in Malta

Sandstone classification schemes vary widely, but most geologists have adopted the Dott scheme, which uses the relative abundance of quartz, feldspar, and lithic framework grains and the abundance of a muddy matrix between the larger grains.

*Composition of framework grains*

> The relative abundance of sand-sized framework grains determines the first word in a sandstone name. Naming depends on the dominance of the three most abundant components quartz, feldspar, or the lithic fragments that originated from other rocks. All other minerals are considered accessories and not used in the naming of the rock, regardless of abundance.
>
> - Quartz sandstones have >90% quartz grains
>
> - Feldspathic sandstones have <90% quartz grains and more feldspar grains than lithic grains
>
> - Lithic sandstones have <90% quartz grains and more lithic grains than feldspar grains

*Abundance of muddy matrix material between sand grains*

When sand-sized particles are deposited, the space between the grains either remains open or is filled with mud (silt and/or clay sized particle).

- "Clean" sandstones with open pore space (that may later be filled with matrix material) are called arenites.

- Muddy sandstones with abundant (>10%) muddy matrix are called wackes.

Six sandstone names are possible using the descriptors for grain composition (quartz-, feldspathic-, and lithic-) and the amount of matrix (wacke or arenite). For example, a quartz arenite would be composed of mostly (>90%) quartz grains and have little or no clayey matrix between the grains, a lithic wacke would have abundant lithic grains and abundant muddy matrix, etc.

Although the Dott classification scheme is widely used by sedimentologists, common names like greywacke, arkose, and quartz sandstone are still widely used by non-specialists and in popular literature.

## Mudrocks

Lower Antelope Canyon was carved out of the surrounding sandstone by both mechanical weathering and chemical weathering. Wind, sand, and water from flash flooding are the primary weathering agents.

Mudrocks are sedimentary rocks composed of at least 50% silt- and clay-sized particles. These relatively fine-grained particles are commonly transported by turbulent flow in water or air, and deposited as the flow calms and the particles settle out of suspension.

Most authors presently use the term "mudrock" to refer to all rocks composed dominantly of mud. Mudrocks can be divided into siltstones, composed dominantly of silt-sized particles; mudstones with subequal mixture of silt- and clay-sized particles; and claystones, composed mostly of clay-sized particles. Most authors use "shale" as a term for a fissile mudrock (regardless of grain size) although some older literature uses the term "shale" as a synonym for mudrock.

# Biochemical Sedimentary Rocks

Outcrop of Ordovician oil shale (kukersite), northern Estonia

Biochemical sedimentary rocks are created when organisms use materials dissolved in air or water to build their tissue. Examples include:

- Most types of limestone are formed from the calcareous skeletons of organisms such as corals, mollusks, and foraminifera.

- Coal, formed from plants that have removed carbon from the atmosphere and combined it with other elements to build their tissue.

- Deposits of chert formed from the accumulation of siliceous skeletons of microscopic organisms such as radiolaria and diatoms.

# Chemical Sedimentary Rocks

Chemical sedimentary rock forms when mineral constituents in solution become supersaturated and inorganically precipitate. Common chemical sedimentary rocks include oolitic limestone and rocks composed of evaporite minerals, such as halite (rock salt), sylvite, barite and gypsum.

# "Other" Sedimentary Rocks

This fourth miscellaneous category includes rocks formed by Pyroclastic flows, impact breccias, volcanic breccias, and other relatively uncommon processes.

# Compositional Classification Schemes

Alternatively, sedimentary rocks can be subdivided into compositional groups based on their mineralogy:

- Siliciclastic sedimentary rocks, are dominantly composed of silicate minerals. The sediment that makes up these rocks was transported as bed load, suspended load, or by sediment gravity flows. Siliciclastic sedimentary rocks are subdivided into conglomerates and breccias, sandstone, and mudrocks.

- Carbonate sedimentary rocks are composed of calcite (rhombohedral $CaCO_3$), aragonite (orthorhombic $CaCO_3$), dolomite ($CaMg(CO_3)_2$), and other carbonate minerals based on the $CO_3^{2-}$ ion. Common examples include limestone and dolostone.

- Evaporite sedimentary rocks are composed of minerals formed from the evaporation of water. The most common evaporite minerals are carbonates (calcite and others based on $CO_3^{2-}$), chlorides (halite and others built on $Cl^-$), and sulfates (gypsum and others built on $SO_4^{2-}$). Evaporite rocks commonly include abundant halite (rock salt), gypsum, and anhydrite.

- Organic-rich sedimentary rocks have significant amounts of organic material, generally in excess of 3% total organic carbon. Common examples include coal, oil shale as well as source rocks for oil and natural gas.

- Siliceous sedimentary rocks are almost entirely composed of silica ($SiO_2$), typically as chert, opal, chalcedony or other microcrystalline forms.

- Iron-rich sedimentary rocks are composed of >15% iron; the most common forms are banded iron formations and ironstones.

- Phosphatic sedimentary rocks are composed of phosphate minerals and contain more than 6.5% phosphorus; examples include deposits of phosphate nodules, bone beds, and phosphatic mudrocks.

## Deposition and Transformation

## Sediment Transport and Deposition

Cross-bedding and scour in a fine sandstone;
the Logan Formation (Mississippian) of Jackson County, Ohio

Sedimentary rocks are formed when sediment is deposited out of air, ice, wind, gravity, or water flows carrying the particles in suspension. This sediment is often formed when weathering and erosion break down a rock into loose material in a source area. The material is then transported from the source area to the deposition area. The type of sediment transported depends on the geology of the hinterland (the source area of the sediment). However, some sedimentary rocks, such as evaporites, are composed of material that form at the place of deposition. The nature of a sedimentary rock, therefore, not only depends on the sediment supply, but also on the sedimentary depositional environment in which it formed.

## Transformation (Diagenesis)

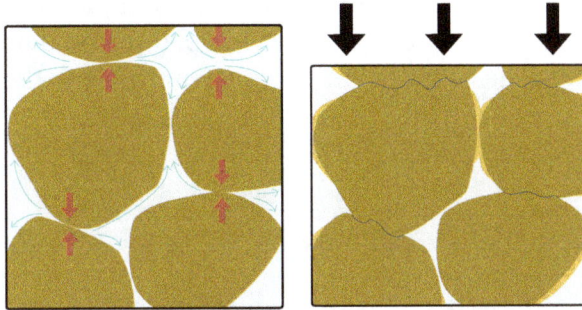

Pressure solution at work in a clastic rock. While material dissolves at places where grains are in contact, that material may recrystallize from the solution and act as cement in open pore spaces. As a result, there is a net flow of material from areas under high stress to those under low stress, producing a sedimentary rock that is more compact and harder. Loose sand can become sandstone in this way.

The term diagenesis is used to describe all the chemical, physical, and biological changes, exclusive of surface weathering, undergone by a sediment after its initial deposition. Some of those processes cause the sediment to consolidate into a compact, solid substance from the originally loose material. Young sedimentary rocks, especially those of Quaternary age (the most recent period of the geologic time scale) are often still unconsolidated. As sediment deposition builds up, the overburden (lithostatic) pressure rises, and a process known as lithification takes place.

Sedimentary rocks are often saturated with seawater or groundwater, in which minerals can dissolve, or from which minerals can precipitate. Precipitating minerals reduce the pore space in a rock, a process called cementation. Due to the decrease in pore space, the original connate fluids are expelled. The precipitated minerals form a cement and make the rock more compact and competent. In this way, loose clasts in a sedimentary rock can become "glued" together.

When sedimentation continues, an older rock layer becomes buried deeper as a result. The lithostatic pressure in the rock increases due to the weight of the overlying sediment. This causes compaction, a process in which grains mechanically reorganize. Compaction is, for example, an important diagenetic process in clay, which can initially consist of 60% water. During compaction, this interstitial water is pressed out of pore spaces. Compaction can also be the result of dissolution of grains by pressure solution. The dissolved material precipitates again in open pore spaces, which means there is a net flow of material into the pores. However, in some cases, a certain mineral dissolves and does not precipitate again. This process, called leaching, increases pore space in the rock.

Some biochemical processes, like the activity of bacteria, can affect minerals in a rock and are therefore seen as part of diagenesis. Fungi and plants (by their roots) and various other organisms that live beneath the surface can also influence diagenesis.

Burial of rocks due to ongoing sedimentation leads to increased pressure and temperature, which stimulates certain chemical reactions. An example is the reactions by which organic material becomes lignite or coal. When temperature and pressure increase still further, the realm of diagenesis makes way for metamorphism, the process that forms metamorphic rock.

## Properties

### Color

The color of a sedimentary rock is often mostly determined by iron, an element with two major oxides: iron(II) oxide and iron(III) oxide. Iron(II) oxide (FeO) only forms under low oxygen (anoxic) circumstances and gives the rock a grey or greenish colour. Iron(III) oxide ($Fe_2O_3$) in a richer oxygen environment is often found in the form of the mineral hematite and gives the rock a reddish to brownish colour. In arid continental climates rocks are in direct contact with the atmosphere, and oxidation is an important process, giving the rock a red or orange colour. Thick sequences of red sedimentary rocks formed in arid climates are called red beds. However, a red colour does not necessarily mean the rock formed in a continental environment or arid climate.

A piece of a banded iron formation, a type of rock that consists of alternating layers with iron(III) oxide (red) and iron(II) oxide (grey). BIFs were mostly formed during the Precambrian, when the atmosphere was not yet rich in oxygen. Moories Group, Barberton Greenstone Belt, South Africa

The presence of organic material can colour a rock black or grey. Organic material is formed from dead organisms, mostly plants. Normally, such material eventually decays by oxidation or bacterial activity. Under anoxic circumstances, however, organic material cannot decay and leaves a dark sediment, rich in organic material. This can, for example, occur at the bottom of deep seas and lakes. There is little water mixing in such environments; as a result, oxygen from surface water is not brought down, and the deposited sediment is normally a fine dark clay. Dark rocks, rich in organic material, are therefore often shales.

## Texture

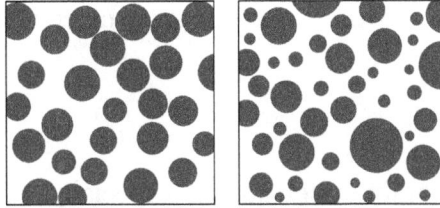

Diagram showing well-sorted (left) and poorly sorted (right) grains

The size, form and orientation of clasts (the original pieces of rock) in a sediment is called its texture. The texture is a small-scale property of a rock, but determines many of its large-scale properties, such as the density, porosity or permeability.

The 3D orientation of the clasts is called the fabric of the rock. Between the clasts, the rock can be composed of a matrix (a cement) that consists of crystals of one or more precipitated minerals. The size and form of clasts can be used to determine the velocity and direction of current in the sedimentary environment that moved the clasts from their origin; fine, calcareous mud only settles in quiet water while gravel and larger clasts are moved only by rapidly moving water. The grain size of a rock is usually expressed with the Wentworth scale, though alternative scales are sometimes used. The grain size can be expressed as a diameter or a volume, and is always an average value – a rock is composed of clasts with different sizes. The statistical distribution of grain sizes is different for different rock types and is described in a property called the sorting of the rock. When all clasts are more or less of the same size, the rock is called 'well-sorted', and when there is a large spread in grain size, the rock is called 'poorly sorted'.

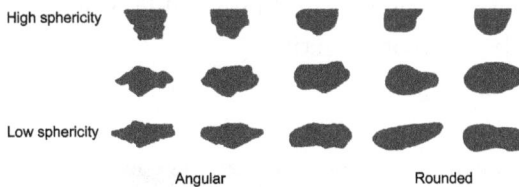

Diagram showing the rounding and sphericity of grains

The form of the clasts can reflect the origin of the rock.

Coquina, a rock composed of clasts of broken shells, can only form in energetic water. The form of a clast can be described by using four parameters:

- *Surface texture* describes the amount of small-scale relief of the surface of a grain that is too small to influence the general shape.

- *rounding* describes the general smoothness of the shape of a grain.

- 'Sphericity' describes the degree to which the grain approaches a sphere.

- 'Grain form' describes the three dimensional shape of the grain.

Chemical sedimentary rocks have a non-clastic texture, consisting entirely of crystals. To describe such a texture, only the average size of the crystals and the fabric are necessary.

## Mineralogy

Most sedimentary rocks contain either quartz (especially siliciclastic rocks) or calcite (especially carbonate rocks). In contrast to igneous and metamorphic rocks, a sedimentary rock usually contains very few different major minerals. However, the origin of the minerals in a sedimentary rock is often more complex than in an igneous rock. Minerals in a sedimentary rock can have formed by precipitation during sedimentation or by diagenesis. In the second case, the mineral precipitate can have grown over an older generation of cement. A complex diagenetic history can be studied by optical mineralogy, using a petrographic microscope.

Carbonate rocks dominantly consist of carbonate minerals such as calcite, aragonite or dolomite. Both the cement and the clasts (including fossils and ooids) of a carbonate sedimentary rock can consist of carbonate minerals. The mineralogy of a clastic rock is determined by the material supplied by the source area, the manner of its transport to the place of deposition and the stability of that particular mineral. The resistance of rock-forming minerals to weathering is expressed by Bowen's reaction series. In this series, quartz is the most stable, followed by feldspar, micas, and finally other less stable minerals that are only present when little weathering has occurred. The amount of weathering depends mainly on the distance to the source area, the local climate and the time it took for the sediment to be transported to the point where it is deposited. In most sedimentary rocks, mica, feldspar and less stable minerals have been reduced to clay minerals like kaolinite, illite or smectite.

## Fossils

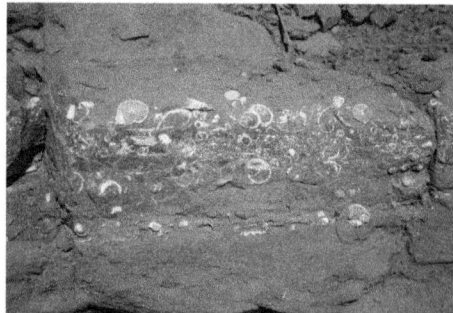

Fossil-rich layers in a sedimentary rock, Año Nuevo State Reserve, California

Among the three major types of rock, fossils are most commonly found in sedimentary rock. Unlike most igneous and metamorphic rocks, sedimentary rocks form at temperatures and pressures that do not destroy fossil remnants. Often these fossils may only be visible under magnification.

Dead organisms in nature are usually quickly removed by scavengers, bacteria, rotting and erosion, but sedimentation can contribute to exceptional circumstances where these natural processes are unable to work, causing fossilisation. The chance of fossilisation is higher when the sedimentation rate is high (so that a carcass is quickly buried), in anoxic environments (where little bacterial activity occurs) or when the organism had a particularly hard skeleton. Larger, well-preserved fossils are relatively rare.

Burrows in a turbidite, made by crustaceans, San Vincente Formation (early Eocene) of the Ainsa Basin, southern foreland of the Pyrenees

Fossils can be both the direct remains or imprints of organisms and their skeletons. Most commonly preserved are the harder parts of organisms such as bones, shells, and the woody tissue of plants. Soft tissue has a much smaller chance of being fossilized, and the preservation of soft tissue of animals older than 40 million years is very rare. Imprints of organisms made while they were still alive are called trace fossils, examples of which are burrows, footprints, etc.

As a part of a sedimentary or metamorphic rock, fossils undergo the same diagenetic processes as does the containing rock. A shell consisting of calcite can, for example, dissolve while a cement of silica then fills the cavity. In the same way, precipitating minerals can fill cavities formerly occupied by blood vessels, vascular tissue or other soft tissues. This preserves the form of the organism but changes the chemical composition, a process called permineralization. The most common minerals involved in permineralization are cements of carbonates (especially calcite), forms of amorphous silica (chalcedony, flint, chert) and pyrite. In the case of silica cements, the process is called lithification.

At high pressure and temperature, the organic material of a dead organism undergoes chemical reactions in which volatiles such as water and carbon dioxide are expulsed.

The fossil, in the end, consists of a thin layer of pure carbon or its mineralized form, graphite. This form of fossilisation is called carbonisation. It is particularly important for plant fossils. The same process is responsible for the formation of fossil fuels like lignite or coal.

## Primary Sedimentary Structures

Structures in sedimentary rocks can be divided into 'primary' structures (formed during deposition) and 'secondary' structures (formed after deposition). Unlike textures, structures are always large-scale features that can easily be studied in the field. Sedimentary structures can indicate something about the sedimentary environment or can serve to tell which side originally faced up where tectonics have tilted or overturned sedimentary layers.

Cross-bedding in a fluviatile sandstone, Middle Old Red Sandstone (Devonian) on Bressay, Shetland Islands

A flute cast, a type of sole marking, from the Book Cliffs of Utah

Sedimentary rocks are laid down in layers called beds or strata. A bed is defined as a layer of rock that has a uniform lithology and texture. Beds form by the deposition of layers of sediment on top of each other. The sequence of beds that characterizes sedimentary rocks is called bedding. Single beds can be a couple of centimetres to several meters thick. Finer, less pronounced layers are called laminae, and the structure a lamina forms in a rock is called lamination. Laminae are usually less than a few centimetres thick. Though bedding and lamination are often originally horizontal in nature, this is not always the case. In some environments, beds are deposited at a (usually small) angle. Sometimes multiple sets of layers with different orientations exist in the same rock, a structure called cross-bedding. Cross-bedding forms when small-scale erosion occurs during deposition, cutting off part of the beds. Newer beds then form at an angle to older ones.

Ripple marks formed by a current in a sandstone that was later tilted (Haßberge, Bavaria)

The opposite of cross-bedding is parallel lamination, where all sedimentary layering is parallel. Differences in laminations are generally caused by cyclic changes in the sediment supply, caused, for example, by seasonal changes in rainfall, temperature or biochemical activity. Laminae that represent seasonal changes (similar to tree rings) are called varves. Any sedimentary rock composed of millimeter or finer scale layers can be named with the general term *laminite*. When sedimentary rocks have no lamination at all, their structural character is called massive bedding.

Graded bedding is a structure where beds with a smaller grain size occur on top of beds with larger grains. This structure forms when fast flowing water stops flowing. Larger, heavier clasts in suspension settle first, then smaller clasts. Although graded bedding can form in many different environments, it is a characteristic of turbidity currents.

The surface of a particular bed, called the bedform, can be indicative of a particular sedimentary environment, too. Examples of bed forms include dunes and ripple marks. Sole markings, such as tool marks and flute casts, are groves dug into a sedimentary layer that are preserved. These are often elongated structures and can be used to establish the direction of the flow during deposition.

Ripple marks also form in flowing water. There are two types of ripples: symmetric and asymmetric. Environments where the current is in one direction, such as rivers, produce asymmetric ripples. The longer flank of such ripples is on the upstream side of the current. Symmetric wave ripples occur in environments where currents reverse directions, such as tidal flats.

Mudcracks are a bed form caused by the dehydration of sediment that occasionally comes above the water surface. Such structures are commonly found at tidal flats or point bars along rivers.

## Secondary Sedimentary Structures

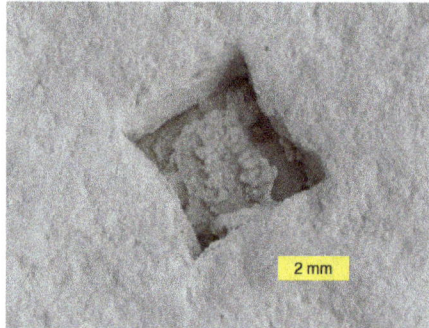

Halite crystal mold in dolomite, Paadla Formation (Silurian), Saaremaa, Estonia

Secondary sedimentary structures are those which formed after deposition. Such structures form by chemical, physical and biological processes within the sediment. They can be indicators of circumstances after deposition. Some can be used as way up criteria.

Organic materials in a sediment can leave more traces than just fossils. Preserved tracks and burrows are examples of trace fossils (also called ichnofossils). Such traces are relatively rare. Most trace fossils are burrows of molluscs or arthropods. This burrowing is called bioturbation by sedimentologists. It can be a valuable indicator of the biological and ecological environment that existed after the sediment was deposited. On the other hand, the burrowing activity of organisms can destroy other (primary) structures in the sediment, making a reconstruction more difficult.

Chert concretions in chalk, Middle Lefkara Formation (upper Paleocene to middle Eocene), Cyprus

Secondary structures can also form by diagenesis or the formation of a soil (pedogenesis) when a sediment is exposed above the water level. An example of a diagenetic structure common in carbonate rocks is a stylolite. Stylolites are irregular planes where material was dissolved into the pore fluids in the rock. This can result in the precipitation of a certain chemical species producing colouring and staining of the rock, or the formation of concretions. Concretions are roughly concentric bodies with a different composition from the host rock. Their formation can be the result of localized precipitation due to small differences in composition or porosity of the host rock, such as around fossils, inside burrows or around plant roots. In carbonate based rocks such as limestone or chalk, chert or flint concretions are common, while terrestrial sandstones can have iron concretions. Calcite concretions in clay are called septarian concretions.

After deposition, physical processes can deform the sediment, producing a third class of secondary structures. Density contrasts between different sedimentary layers, such as between sand and clay, can result in flame structures or load casts, formed by inverted diapirism. While the clastic bed is still fluid, diapirism can cause a denser upper layer to sink into a lower layer. Sometimes, density contrasts can result or grow when one of the lithologies dehydrates. Clay can be easily compressed as a result of dehydration, while sand retains the same volume and becomes relatively less dense. On the other hand, when the pore fluid pressure in a sand layer surpasses a critical point, the sand can break through overlying clay layers and flow through, forming discordant bodies of sedimentary rock called sedimentary dykes. The same process can form mud volcanoes on the surface where they broke through upper layers.

Sedimentary dykes can also be formed in a cold climate where the soil is permanently frozen during a large part of the year. Frost weathering can form cracks in the soil that fill with rubble from above. Such structures can be used as climate indicators as well as way up structures.

Density contrasts can also cause small-scale faulting, even while sedimentation progresses (synchronous-sedimentary faulting). Such faulting can also occur when large masses of non-lithified sediment are deposited on a slope, such as at the front side of a delta or the continental slope. Instabilities in such sediments can result in the deposited material to slump, producing fissures and folding. The resulting structures in the rock are syn-sedimentary folds and faults, which can be difficult to distinguish from folds and faults formed by tectonic forces acting on lithified rocks.

## Sedimentary Environments

The setting in which a sedimentary rock forms is called the sedimentary environment. Every environment has a characteristic combination of geologic processes and circumstances. The type of sediment that is deposited is not only dependent on the sediment that is transported to a place, but also on the environment itself.

A marine environment means that the rock was formed in a sea or ocean. Often, a distinction is made between deep and shallow marine environments. Deep marine usually refers to environments more than 200 m below the water surface. Shallow marine environments exist adjacent to coastlines and can extend to the boundaries of the continental shelf. The water movements in such environments have a generally higher energy than that in deep environments, as wave activity diminishes with depth. This means that coarser sediment particles can be transported and the deposited sediment can be coarser than in deeper environments. When the sediment is transported from the continent, an alternation of sand, clay and silt is deposited. When the continent is far away, the amount of such sediment deposited may be small, and biochemical processes dominate the type of rock that forms. Especially in warm climates, shallow marine environments far offshore mainly see deposition of carbonate rocks. The shallow, warm water is an ideal habitat for many small organisms that build carbonate skeletons. When these organisms die, their skeletons sink to the bottom, forming a thick layer of calcareous mud that may lithify into limestone. Warm shallow marine environments also are ideal environments for coral reefs, where the sediment consists mainly of the calcareous skeletons of larger organisms.

In deep marine environments, the water current working the sea bottom is small. Only fine particles can be transported to such places. Typically sediments depositing on the ocean floor are fine clay or small skeletons of micro-organisms. At 4 km depth, the solubility of carbonates increases dramatically (the depth zone where this happens is called the lysocline). Calcareous sediment that sinks below the lysocline dissolves; as a result, no limestone can be formed below this depth. Skeletons of micro-organisms formed of silica (such as radiolarians) are not as soluble and still deposit. An example of a rock formed of silica skeletons is radiolarite. When the bottom of the sea has a small inclination, for example at the continental slopes, the sedimentary cover can become unstable, causing turbidity currents. Turbidity currents are sudden disturbances of the normally quite deep marine environment and can cause the geologically speaking instantaneous deposition of large amounts of sediment, such as sand and silt. The rock sequence formed by a turbidity current is called a turbidite.

The coast is an environment dominated by wave action. At a beach, dominantly denser sediment such as sand or gravel, often mingled with shell fragments, is deposited, while the silt and clay sized material is kept in mechanical suspension. Tidal flats and shoals are places that sometimes dry because of the tide. They are often cross-cut by gullies, where the current is strong and the grain size of the deposited sediment is larger. Where rivers enter the body of water, either on a sea or lake coast, deltas can form. These are large accumulations of sediment transported from the continent to places in front of the mouth of the river. Deltas are dominantly composed of clastic sediment (in contrast to chemical).

A sedimentary rock formed on land has a continental sedimentary environment. Examples of continental environments are lagoons, lakes, swamps, floodplains and alluvial fans. In the quiet water of swamps, lakes and lagoons, fine sediment is deposited, mingled with organic material from dead plants and animals. In rivers, the energy of the water is much greater and can transport heavier clastic material. Besides transport by water, sediment can in continental environments also be transported by wind or glaciers. Sediment transported by wind is called aeolian and is always very well sorted, while sediment transported by a glacier is called glacial till and is characterized by very poor sorting.

Aeolian deposits can be quite striking. The depositional environment of the Touchet Formation, located in the Northwestern United States, had intervening periods of aridity which resulted in a series of rhythmite layers. Erosional cracks were later infilled with layers of soil material, especially from aeolian processes. The infilled sections formed vertical inclusions in the horizontally deposited layers of the Touchet Formation, and thus provided evidence of the events that intervened over time among the forty-one layers that were deposited.

## Sedimentary Facies

Sedimentary environments usually exist alongside each other in certain natural successions. A beach, where sand and gravel is deposited, is usually bounded by a deeper marine environment a little offshore, where finer sediments are deposited at the same time. Behind the beach, there can be dunes (where the dominant deposition is well sorted sand) or a lagoon (where fine clay and organic material is deposited). Every sedimentary environment has its own characteristic deposits. The typical rock formed in a certain environment is called its sedimentary facies. When sedimentary strata accumulate through time, the environment can shift, forming a change in facies in the subsurface at one location. On the other hand, when a rock layer with a certain age is followed laterally, the lithology (the type of rock) and facies eventually change.

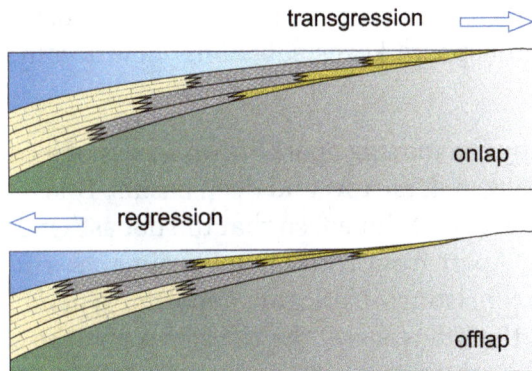

Shifting sedimentary facies in the case of transgression (above) and regression of the sea (below)

Facies can be distinguished in a number of ways: the most common are by the lithology (for example: limestone, siltstone or sandstone) or by fossil content. Coral, for example, only lives in warm and shallow marine environments and fossils of coral are thus typical for shallow marine facies. Facies determined by lithology are called lithofacies; facies determined by fossils are biofacies.

Sedimentary environments can shift their geographical positions through time. Coastlines can shift in the direction of the sea when the sea level drops, when the surface rises due to tectonic forces in the Earth's crust or when a river forms a large delta. In the subsurface, such geographic shifts of sedimentary environments of the past are recorded in shifts in sedimentary facies. This means that sedimentary facies can change either parallel or perpendicular to an imaginary layer of rock with a fixed age, a phenomenon described by Walther's Law.

The situation in which coastlines move in the direction of the continent is called transgression. In the case of transgression, deeper marine facies are deposited over shallower facies, a succession called onlap. Regression is the situation in which a coastline moves in the direction of the sea. With regression, shallower facies are deposited on top of deeper facies, a situation called offlap.

The facies of all rocks of a certain age can be plotted on a map to give an overview of the palaeogeography. A sequence of maps for different ages can give an insight in the development of the regional geography.

## Sedimentary Basins

Places where large-scale sedimentation takes place are called sedimentary basins. The amount of sediment that can be deposited in a basin depends on the depth of the basin, the so-called accommodation space. The depth, shape and size of a basin depend on tectonics, movements within the Earth's lithosphere. Where the lithosphere moves upward (tectonic uplift), land eventually rises above sea level, so that and erosion removes material, and the area becomes a source for new sediment. Where the lithosphere moves downward (tectonic subsidence), a basin forms and sedimentation can take place. When the lithosphere keeps subsiding, new accommodation space keeps being created.

A type of basin formed by the moving apart of two pieces of a continent is called a rift basin. Rift basins are elongated, narrow and deep basins. Due to divergent movement, the lithosphere is stretched and thinned, so that the hot asthenosphere rises and heats the overlying rift basin. Apart from continental sediments, rift basins normally also have part of their infill consisting of volcanic deposits. When the basin grows due to continued stretching of the lithosphere, the rift grows and the sea can enter, forming marine deposits.

When a piece of lithosphere that was heated and stretched cools again, its density rises, causing isostatic subsidence. If this subsidence continues long enough, the basin is called a sag basin. Examples of sag basins are the regions along passive continental margins, but sag basins can also be found in the interior of continents. In sag basins, the extra weight of the newly deposited sediments is enough to keep the subsidence going in a vicious circle. The total thickness of the sedimentary infill in a sag basins can thus exceed 10 km.

A third type of basin exists along convergent plate boundaries – places where one tectonic plate moves under another into the asthenosphere. The subducting plate bends and forms a fore-arc basin in front of the overriding plate—an elongated, deep asymmetric basin. Fore-arc basins are filled with deep marine deposits and thick sequences of turbidites. Such infill is called flysch. When the convergent movement of the two plates results in continental collision, the basin becomes shallower and develops into a foreland basin. At the same time, tectonic uplift forms a mountain belt in the overriding plate, from which large amounts of material are eroded and transported to the basin. Such erosional material of a growing mountain chain is called molasse and has either a shallow marine or a continental facies.

At the same time, the growing weight of the mountain belt can cause isostatic subsidence in the area of the overriding plate on the other side to the mountain belt. The basin type resulting from this subsidence is called a back-arc basin and is usually filled by shallow marine deposits and molasse.

Cyclic alternation of competent and less competent beds
in the Blue Lias at Lyme Regis, southern England

## Influence of Astronomical Cycles

In many cases facies changes and other lithological features in sequences of sedimentary rock have a cyclic nature. This cyclic nature was caused by cyclic changes in sediment supply and the sedimentary environment. Most of these cyclic changes are caused by astronomic cycles. Short astronomic cycles can be the difference between the tides or the spring tide every two weeks. On a larger time-scale, cyclic changes in climate and sea level are caused by Milankovitch cycles: cyclic changes in the orientation and/or position of the Earth's rotational axis and orbit around the Sun. There are a number of Milankovitch cycles known, lasting between 10,000 and 200,000 years.

Relatively small changes in the orientation of the Earth's axis or length of the seasons can be a major influence on the Earth's climate. An example are the ice ages of the past 2.6 million years (the Quaternary period), which are assumed to have been caused by astronomic cycles. Climate change can influence the global sea level (and thus the amount of accommodation space in sedimentary basins) and sediment supply from a certain region. Eventually, small changes in astronomic parameters can cause large changes in sedimentary environment and sedimentation.

## Sedimentation Rates

The rate at which sediment is deposited differs depending on the location. A channel in a tidal flat can see the deposition of a few metres of sediment in one day, while on the deep ocean floor each year only a few millimetres of sediment accumulate. A distinction can be made between normal sedimentation and sedimentation caused by catastrophic processes. The latter category includes all kinds of sudden exceptional processes like mass movements, rock slides or flooding. Catastrophic processes can see the sudden deposition of a large amount of sediment at once. In some sedimentary environments, most of the total column of sedimentary rock was formed by catastrophic processes, even though the environment is usually a quiet place. Other sedimentary environments are dominated by normal, ongoing sedimentation.

In many cases, sedimentation occurs slowly. In a desert, for example, the wind deposits siliciclastic material (sand or silt) in some spots, or catastrophic flooding of a wadi may cause sudden deposits of large quantities of detrital material, but in most places eolian erosion dominates. The amount of sedimentary rock that forms is not only dependent on the amount of supplied material, but also on how well the material consolidates. Erosion removes most deposited sediment shortly after deposition.

## Stratigraphy

Picture from Glen Canyon National Recreation Area, Utah.

The Permian through Jurassic stratigraphy of the Colorado Plateau area of southeastern Utah that makes up much of the famous prominent rock formations in protected areas such as Capitol Reef National Park and Canyonlands National Park. From top to bottom: Rounded tan domes of the Navajo Sandstone, layered red Kayenta Formation, cliff-forming, vertically jointed, red Wingate Sandstone, slope-forming, purplish Chinle Formation, layered, lighter-red Moenkopi Formation, and white, layered Cutler Formation sandstone.

That new rock layers are above older rock layers is stated in the principle of superposition. There are usually some gaps in the sequence called unconformities. These represent periods where no new sediments were laid down, or when earlier sedimentary layers were raised above sea level and eroded away.

Sedimentary rocks contain important information about the history of the Earth. They contain fossils, the preserved remains of ancient plants and animals. Coal is considered a type of sedimentary rock. The composition of sediments provides us with clues as to the original rock. Differences between successive layers indicate changes to the environment over time. Sedimentary rocks can contain fossils because, unlike most igneous and metamorphic rocks, they form at temperatures and pressures that do not destroy fossil remains.

## Clastic Rocks

Clastic rocks comprise siliciclastic sediments which are made up of physically deposited particles such as grains of quartz and feldspar derived from weathered pre-existing rocks (the term 'clastic' is derived from the greek word *klastos*, meaning "broken"). These sediments are laid down by geological agents like water, wind and ice. The most abundant silicate minerals in siliciclastics sedimentary rocks are quartz, feldspar and clay minerals. Clay minerals are formed by weathering and alteration of pre-existing silicate minerals, such as feldspar. Some dark minerals like pyroxene and amphiboles, micas and garnet may also be present. Sediments are the precursors of sedimentary rocks that are found at Earth's surface as layers of loose particles, such as sand, silt, and the shells of organisms. These particles originate in the processes of weathering and erosion. The loose grains of sediment transform into sedimentary rock by following five steps:

- *Weathering* refers to the entire chemical, physical and biological processes that break up and decay rocks into fragments and dissolved substances of various sizes. These particles are then transported by erosion, the set of processes that loosen soil and rock - and move them downhill or down stream to a place where they are deposited as layers of sediments.

- *Erosion* refers to the combination of processes that separate rock or regolith such as abrasion, plucking caused by moving air, water or ice.

- *Transportation* can occur by gravity, wind, water or ice. They can carry sediments. The ability of a medium to carry sediment depends on its viscosity and velocity.

- *Deposition* is the process by which sediments (a) settles out of transporting medium due to decrease in velocity or (b) precipitate from a solution due to saturation or change in temperature/pressure, the medium is no longer sediment carry.

- *Lithification* is the transformation of the loose sediment into solid rock. During lithification the sediments accumulate in layers, compress under their own weight and/or what buries them and form a hardened mass.

# Sedimentary Clastic Rocks

Claystone from Montana

Clastic sedimentary rocks are rocks composed predominantly of broken pieces or *clasts* of older weathered and eroded rocks. Clastic sediments or sedimentary rocks are classified based on grain size, clast and cementing material (matrix) composition, and texture. The classification factors are often useful in determining a sample's environment of deposition. An example clastic environment would be a river system in which the full range of grains being transported by the moving water consist of pieces eroded from solid rock upstream.

Grain size varies from clay in shales and claystones; through silt in siltstones; sand in sandstones; and gravel, cobble, to boulder sized fragments in conglomerates and breccias. The Krumbein phi ($\varphi$) scale numerically orders these terms in a logarithmic size scale.

## Siliciclastic Sedimentary Rocks

*Siliciclastic* rocks are clastic noncarbonate rocks that are composed almost exclusively of silicon, either as forms of quartz or as silicates.

## Composition

The composition of siliciclastic sedimentary rocks includes the chemical and mineralogical components of the framework as well as the cementing material that make up these rocks. Boggs divides them into four categories; major minerals, accessory minerals, rock fragments, and chemical sediments.

Major minerals can be categorized into subdivisions based on their resistance to chemical decomposition. Those that possess a great resistance to decomposition are categorized as stable, while those that do not are considered less stable. The most common stable mineral in siliciclastic sedimentary rocks is quartz ($SiO_2$). Quartz makes up approximately 65 percent of framework grains present in sandstones and about

30 percent of minerals in the average shale. Less stable minerals present in this type of rocks are feldspars, including both potassium and plagioclase feldspars. Feldspars comprise a considerably lesser portion of framework grains and minerals. They only make up about 15 percent of framework grains in sandstones and 5% of minerals in shales. Clay mineral groups are mostly present in mudrocks (comprising more than 60% of the minerals) but can be found in other siliciclastic sedimentary rocks at considerably lower levels.

Accessory minerals are associated with those whose presence in the rock are not directly important to the classification of the specimen. These generally occur in smaller amounts in comparison to the quartz, and feldspars. Furthermore, those that do occur are generally heavy minerals or coarse grained micas (both Muscovite and Biotite).

Rock fragments also occur in the composition of siliciclastic sedimentary rocks and are responsible for about 10–15 percent of the composition of sandstone. They generally make up most of the gravel size particles in conglomerates but contribute only a very small amount to the composition of mudrocks. Though they sometimes are, rock fragments are not always sedimentary in origin. They can also be metamorphic or igneous.

Chemical cements vary in abundance but are predominantly found in sandstones. The two major types, are silicate based and carbonate based. The majority of silica cements are composed of quartz but can include, chert, opal, feldspars and zeolites.

Composition includes the chemical and mineralogic make-up of the single or varied fragments and the cementing material (matrix) holding the clasts together as a rock. These differences are most commonly used in the framework grains of sandstones. Sandstones rich in quartz are called quartz arenites, those rich in feldspar are called arkoses, and those rich in lithics are called lithic sandstones.

## Classification

Siliciclastic sedimentary rocks are composed of mainly silicate particles derived by the weathering of older rocks and pyroclastic volcanism. While grain size, clast and cementing material (matrix) composition, and texture are important factors when regarding composition, siliciclastic sedimentary rocks are classified according to grain size into three major categories; conglomerates, sandstones, and mudrocks. The term clay is used to classify particles smaller than .0039 millimeters. However, term can also be used to refer to a family of sheet silicate minerals. Silt refers to particles that have a diameter between .062 and .0039 millimeters. The term *mud* is used when clay and silt particles are mixed in the sediment; *mudrock* is the name of the rock created with these sediments. Furthermore, particles that reach diameters between .062 and 2 millimeters fall into the category of sand. When sand is cemented together and lithified it becomes known as sandstone. Any particle that is larger than two millimeters is considered gravel. This category includes pebbles, cobbles and boulders. Like sandstone, when gravels are lithified they are considered conglomerates.

## Conglomerates and Breccias

Conglomerate

Conglomerates are coarse grained rocks dominantly composed of gravel sized particles that are typically held together by a finer grained matrix. These rocks are often subdivided into conglomerates and breccias. The major characteristic that divides these two categories is the amount of rounding. The gravel sized particles that make up conglomerates are well rounded while in breccias they are angular. Conglomerates are common in stratigraphic successions of most, if not all ages but only make up one percent or less, by weight of the total sedimentary rock mass. In terms or origin and depositional mechanisms they are very similar to sandstones. As a result, the two categories often contain the same sedimentary structures.

Breccia. Notice the angular nature of the large clasts

## Sandstones

Sandstones are medium-grained rocks composed of rounded or angular fragments of sand size, that often but not always have a cement uniting them together. These sand-size particles are often quartz but there are a few common categories and a wide variety of classification schemes that classify sandstones based on composition. Classification schemes vary widely, but most geologists have adopted the Dott scheme, which uses the relative abundance of quartz, feldspar, and lithic framework grains and the abundance of muddy matrix between these larger grains.

Sandstone from Lower Antelope Canyon

## Mudrocks

Rocks that are classified as mudrocks are very fine grained. Silt and clay represent at least 50% of the material that mudrocks are composed of. Classification schemes for mudrocks tend to vary but most are based on the grain size of the major constituents. In mudrocks, these are generally silt, and clay.

According to Blatt, Middleton and Murray  mudrocks that are composed mainly of silt particles are classified as siltstones. In turn, rocks that possess clay as the majority particle are called claystones. In geology, a mixture of both silt and clay is called mud. Rocks that possess large amounts of both clay and silt are called mudstones. In some cases the term shale is also used to refer to mudrocks and is still widely accepted by most. However, others have used the term shale to further divide mudrocks based on the percentage of clay constituents. The plate-like shape of clay allows its particles to stack up one on top of another creating laminae or beds. The more clay present in a given specimen, the more laminated a rock is. Shale, in this case, is reserved for mudrocks that are laminated, while mudstone refers those that are not.

Red mudrock

Black Shale

## Diagenesis of Siliciclastic Sedimentary Rocks

Siliciclastic rocks initially form as loosely packed sediment deposits including gravels, sands, and muds. The process of turning loose sediment into hard sedimentary rocks is called lithification. During the process of lithification, sediments undergo physical, chemical and mineralogical changes before becoming rock. The primary physical process in lithification is compaction. As sediment transport and deposition continues, new sediments are deposited atop previously deposited beds burying them. Burial continues and the weight of overlying sediments cause an increase in temperature and pressure. This increase in temperature and pressure causes loose grained sediments become tightly packed reducing porosity, essentially squeezing water out of the sediment. Porosity is further reduced by the precipitation of minerals into the remaining pore spaces. The final stage in the process is diagenesis and will be discussed in detail below.

## Cementation

Cementation is the diagenetic process by which coarse clastic sediments become lithified or consolidated into hard, compact rocks, usually through the deposition or precipitation of minerals in the spaces between the individual grains of sediment. Cementation can occur simultaneously with deposition or at another time. Furthermore, once a sediment is deposited, it becomes subject to cementation through the various stages of diagenesis discussed below.

### Shallow Burial (Eogenesis)

Eogenesis refers to the early stages of diagenesis. This can take place at very shallow depths, ranging from a few meters to tens of meters below the surface. The changes that occur during this diagenetic phase mainly relate to the reworking of the sediments. Compaction and grain repacking, bioturbation, as well as mineralogical changes all occur at varying degrees. Due to the shallow depths, sediments undergo only minor compaction and grain rearrangement during this stage. Organisms rework sediment near the depositional interface by burrowing, crawling, and in some cases sediment ingestion. This process can destroy sedimentary structures that were present upon deposition of the sediment. Structures such as lamination will give way to new structures associated the activity of organisms. Despite being close to the surface, eogenesis does provide conditions for important mineralogical changes to occur. This mainly involves the precipitation of new minerals.

### Mineralogical Changes During Eogenesis

Mineralogical changes that occur during eogenesis as dependent on the environment in which that sediment has been deposited. For example, the formation of pyrite is characteristic of reducing conditions in marine environments. Pyrite can form as cement, or replace organic materials, such as wood fragments. Other important reactions include the formation of chlorite, glauconite, illite and iron oxide (if oxygenated pore water is present). The precipitation of potassium feldspar, quartz overgrowths, and carbonate cements also occurs under marine conditions. In non marine environments oxidizing conditions are almost always prevalent, meaning iron oxides are commonly produced along with kaolin group clay minerals. The precipitation of quartz and calcite cements may also occur in non marine conditions.

## Deep Burial (Mesogenesis)

## Compaction

As sediments are buried deeper, load pressures become greater resulting in tight grain packing and bed thinning. This causes increased pressure between grains thus increasing the solubility of grains. As a result, the partial dissolution of silicate grains occurs. This is called pressure solutions. Chemically speaking, increases in temperature can also cause chemical reaction rates to increase. This increases the solubility of most common minerals (aside from evaporites). Furthermore, beds thin and porosity decreases allowing cementation to occur by the precipitation of silica or carbonate cements into remaining pore space.

In this process minerals crystallize from watery solutions that percolate through the pores between grain of sediment. The cement that is produced may or may not have the same chemical composition as the sediment. In sandstones, framework grains are often cemented by silica or carbonate. The extent of cementation is dependent on the composition of the sediment. For example, in lithic sandstones, cementation is less extensive because pore space between framework grains is filled with a muddy matrix that leaves little space for precipitation to occur. This is often the case for mudrocks as well. As a result of compaction, the clayey sediments comprising mudrocks are relatively impermeable.

## Dissolution

Dissolution of framework silicate grains and previously formed carbonate cement may occur during deep burial. Conditions that encourage this are essentially opposite of those required for cementation. Rock fragments and silicate minerals of low stability, such as plagioclase feldspar, pyroxenes, and amphiboles, may dissolve as a result of increasing burial temperatures and the presence of organic acids in pore waters. The dissolution of frame work grains and cements increases porosity particularly in sandstones.

## Mineral Replacement

This refers to the process whereby one mineral is dissolved and a new mineral fills the space via precipitation. Replacement can be partial or complete. Complete replacement destroys the identity of the original minerals or rock fragments giving a biased view of the original mineralogy of the rock/ Porosity can also be affected by this process. For example, clay minerals tend to fill up pore space and thereby reducing porosity.

## Telogenesis

In the process of burial, it is possible that siliciclastic deposits may subsequently be uplifted as a result of a mountain building event or erosion. When uplift occurs, it exposes buried deposits to a radically new environment. Because the process brings material to or closer to the surface, sediments that undergo uplift are subjected to lower temperatures and pressures as well as slightly acidic rain water. Under these conditions, framework grains and cement are again subjected to dissolution and in turn increasing porosity. On the other hand, telogenesis can also change framework grains to clays, thus reducing porosity. These changes are dependent on the specific conditions that the rock is exposed as well as the composition of the rock and pore waters. Specific pore waters, can cause the further precipitation of carbonate or silica cements. This process can also encourage the process of oxidation on a variety of iron bearing minerals.

## Sedimentary Breccias

Sedimentary breccias are a type of clastic sedimentary rock which are composed of angular to subangular, randomly oriented clasts of other sedimentary rocks. They may form either

1.  in submarine debris flows, avalanches, mud flow or mass flow in an aqueous medium. Technically, turbidites are a form of debris flow deposit and are a fine-grained peripheral deposit to a sedimentary breccia flow.

2.  as angular, poorly sorted, very immature fragments of rocks in a finer grained groundmass which are produced by mass wasting. These are, in essence, lith-ified colluvium. Thick sequences of sedimentary (colluvial) breccias are gener-ally formed next to fault scarps in grabens.

In the field, it may at times be difficult to distinguish between a debris flow sedimen-tary breccia and a colluvial breccia, especially if one is working entirely from drilling information. Sedimentary breccias are an integral host rock for many sedimentary ex-halative deposits.

## Igneous Clastic Rocks

Clastic igneous rocks include pyroclastic volcanic rocks such as tuff, agglomerate and intrusive breccias, as well as some marginal eutaxitic and taxitic intrusive morpholo-gies. Igneous clastic rocks are broken by flow, injection or explosive disruption of solid or semi-solid igneous rocks or lavas.

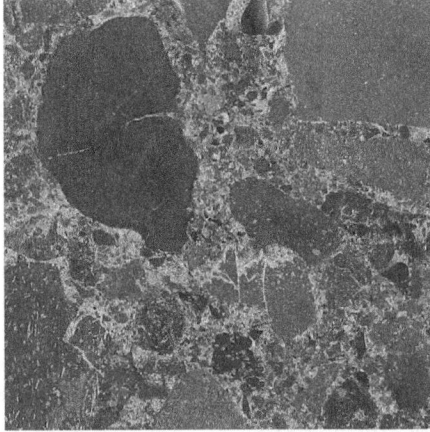

Basalt breccia, green groundmass is composed of epidote

Igneous clastic rocks can be divided into two classes:

1. Broken, fragmental rocks produced by intrusive processes, usually associated with plutons or porphyry stocks

2. Broken, fragmental rocks associated with volcanic eruptions, both of lava and pyroclastic type

## Metamorphic Clastic Rocks

Clastic metamorphic rocks include breccias formed in faults, as well as some proto-mylonite and pseudotachylite. Occasionally, metamorphic rocks can be brecciated via hydrothermal fluids, forming a hydrofracture breccia.

## Hydrothermal Clastic Rocks

Hydrothermal clastic rocks are generally restricted to those formed by hydrofracture, the process by which hydrothermal circulation cracks and brecciates the wall rocks and fills it in with veins. This is particularly prominent in epithermal ore deposits and is associated with alteration zones around many intrusive rocks, especially granites. Many skarn and greisen deposits are associated with hydrothermal breccias.

## Impact Breccias

A fairly rare form of clastic rock may form during meteorite impact. This is composed primarily of ejecta; clasts of country rock, melted rock fragments, tektites (glass ejected from the impact crater) and exotic fragments, including fragments derived from the impactor itself.

Identifying a clastic rock as an impact breccia requires recognising shatter cones, tektites, spherulites, and the morphology of an impact crater, as well as potentially recognizing particular chemical and trace element signatures, especially osmiridium.

## Nonclastic Rocks

Nonclastic rocks consist of the biological and chemical group of sediments that form by the growth of shell masses or cementing together of shells and shell fragments; by the accumulation and subsequent alteration of organic matter from living organisms; or by the precipitation of minerals from water solutions. Calcite is precipitated by marine organisms to form shells or skeletons which form biological sediments when the organisms die. The most abundant minerals of chemical and biological sediments are carbonates such as calcite, the main constituent of limestone.

(a)                                                                                            (b)

(a) Clastic sedimentary rock, coarse and medium grained sandstone,
(b) Non clastic sedimentary rock, limestone

Different kinds of sedimentary rocks are identified on the basis of their mineral composition. According to some estimates, 70% to 85% of all sedimentary rocks on Earth are clastic, whereas 15%-25% are carbonate biochemical or chemical rocks.

Geologists can work backward using evidences provided by a sedimentary rock's mineral content, texture, and physical structure to infer the sources of the sediments from which these rocks were formed and environment of their deposition.

## Metamorphic Rocks

Metamorphic rocks take their name from the Greek words *meta* meaning 'change' (*meta*) and *morphe*, meaning 'form'. A metamorphic rock is one that (a) forms when a pre-existing rocks or protolith; (b) undergoes a solid-state change in response to the modification of its environment. This process of change is called metamorphism. The rocks undergo metamorphism when they are subjected to high temperature and pressures deep within Earth. This results in changes in the mineralogy, texture or chemical composition of any kind of pre-existing rock-igneous, sedimentary or other metamorphic rock-while maintaining its solid form.

The temperatures of metamorphism are below the melting point of the rocks (about

700°C) but high enough (above 250°C) for the rocks to be changed by recrystallisation and chemical reactions. Metamorphism can produce a group of minerals which together make up a metamorphic minerals assemblage. Their texture is defined by the new or re-arrangement of mineral grains. Commonly, the texture results in metamorphic foliation defined by the parallel alignment of platy minerals (such as mica) and/or the presence of alternating light coloured and dark coloured bands. Metamorphic rocks can be grouped into two types (Figure)-foliated, e.g. phyllite, schist and nonfoliated, e.g. quartzite, marble. For example the metamorphism of granite, a rock with randomly oriented crystals can produce a metamorphosed rock like schist showing parallel alignment of platy minerals (such as mica) or gneiss with alternating light coloured and dark coloured bands.

The formation of metamorphic minerals and textures takes place slowly-it may take - millions of years. The most common processes are:

- *Recrystallisation,* which changes the shape and size of grains without changing the identity of the mineral making up the grains.

- *Phase change,* which transforms one mineral into another mineral with the same composition but with a different crystal structure.

- *Metamorphic reaction* or neocrystallisation (from the Greek *neos,* for new) which results in the growth of new mineral crystals that differ from those of the protolith.

- *Pressure solution*, which happens when a wet rock is squeezed more strongly in one direction than in others, producing ions that migrate through the water to precipitate elsewhere.

- *Plastic deformation*, which happens when a rock is squeezed or sheared at elevated temperatures and pressures. Under such conditions minerals behave like soft plastic and change shape without breaking.

(a)                                                              (b)

Metamorphic rocks (a) foliated-schist, (b) nonfoliated-marble

Common minerals of metamorphic rocks are silicate minerals like quartz, feldspar, micas, pyroxenes and amphiboles. Several other silicate minerals like kyanite, andalusite and some varieties of garnet, are good indicators of metamorphism. Calcite is the mineral of marble which is metamorphosed limestone. Similarly quartz is the mineral of quartzite which is metamorphosed sandstone.

## Metamorphic Minerals

Metamorphic minerals are those that form only at the high temperatures and pressures associated with the process of metamorphism. These minerals, known as index minerals, include sillimanite, kyanite, staurolite, andalusite, and some garnet.

Other minerals, such as olivines, pyroxenes, amphiboles, micas, feldspars, and quartz, may be found in metamorphic rocks, but are not necessarily the result of the process of metamorphism. These minerals formed during the crystallization of igneous rocks. They are stable at high temperatures and pressures and may remain chemically unchanged during the metamorphic process. However, all minerals are stable only within certain limits, and the presence of some minerals in metamorphic rocks indicates the approximate temperatures and pressures at which they formed.

The change in the particle size of the rock during the process of metamorphism is called recrystallization. For instance, the small calcite crystals in the sedimentary rock limestone and chalk change into larger crystals in the metamorphic rock marble; in metamorphosed sandstone, recrystallization of the original quartz sand grains results in very compact quartzite, also known as metaquartzite, in which the often larger quartz crystals are interlocked. Both high temperatures and pressures contribute to recrystallization. High temperatures allow the atoms and ions in solid crystals to migrate, thus reorganizing the crystals, while high pressures cause solution of the crystals within the rock at their point of contact.

## Foliation

Folded foliation in a metamorphic rock from near Geirangerfjord, Norway

The layering within metamorphic rocks is called *foliation* (derived from the Latin word *folia*, meaning "leaves"), and it occurs when a rock is being shortened along one axis during recrystallization. This causes the platy or elongated crystals of minerals, such as mica and chlorite, to become rotated such that their long axes are perpendicular to the orientation of shortening. This results in a banded, or foliated rock, with the bands showing the colors of the minerals that formed them.

Textures are separated into foliated and non-foliated categories. Foliated rock is a product of differential stress that deforms the rock in one plane, sometimes creating a plane of cleavage. For example, slate is a foliated metamorphic rock, originating from shale. Non-foliated rock does not have planar patterns of strain.

Rocks that were subjected to uniform pressure from all sides, or those that lack minerals with distinctive growth habits, will not be foliated. Where a rock has been subject to differential stress, the type of foliation that develops depends on the metamorphic grade. For instance, starting with a mudstone, the following sequence develops with increasing temperature: slate is a very fine-grained, foliated metamorphic rock, characteristic of very low grade metamorphism, while phyllite is fine-grained and found in areas of low grade metamorphism, schist is medium to coarse-grained and found in areas of medium grade metamorphism, and gneiss coarse to very coarse-grained, found in areas of high-grade metamorphism. Marble is generally not foliated, which allows its use as a material for sculpture and architecture.

Another important mechanism of metamorphism is that of chemical reactions that occur between minerals without them melting. In the process atoms are exchanged between the minerals, and thus new minerals are formed. Many complex high-temperature reactions may take place, and each mineral assemblage produced provides us with a clue as to the temperatures and pressures at the time of metamorphism.

Metasomatism is the drastic change in the bulk chemical composition of a rock that often occurs during the processes of metamorphism. It is due to the introduction of chemicals from other surrounding rocks. Water may transport these chemicals rapidly over great distances. Because of the role played by water, metamorphic rocks generally contain many elements absent from the original rock, and lack some that originally were present. Still, the introduction of new chemicals is not necessary for recrystallization to occur.

## Types of Metamorphism

## Contact Metamorphism

A contact metamorphic rock made of interlayered calcite and serpentine from the Precambrian of Canada. Once thought to be a pseudofossil called *Eozoön canadense*. Scale in mm.

Contact metamorphism is the name given to the changes that take place when magma is injected into the surrounding solid rock (country rock). The changes that occur are greatest wherever the magma comes into contact with the rock because the temperatures are highest at this boundary and decrease with distance from it. Around the igneous rock that forms from the cooling magma is a metamorphosed zone called a *contact metamorphism aureole*. Aureoles may show all degrees of metamorphism from the contact area to unmetamorphosed (unchanged) country rock some distance away. The formation of important ore minerals may occur by the process of metasomatism at or near the contact zone.

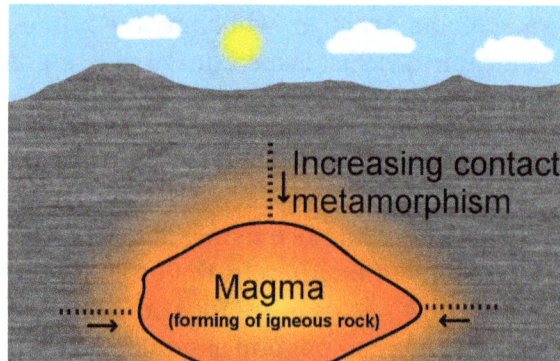

When a rock is contact altered by an igneous intrusion it very frequently becomes more indurated, and more coarsely crystalline. Many altered rocks of this type were formerly called hornstones, and the term *hornfels* is often used by geologists to signify those fine grained, compact, non-foliated products of contact metamorphism. A shale may become a dark argillaceous hornfels, full of tiny plates of brownish biotite; a marl or impure limestone may change to a grey, yellow or greenish lime-silicate-hornfels or siliceous marble, tough and splintery, with abundant augite, garnet, wollastonite and other minerals in which calcite is an important component. A diabase or andesite may become a diabase hornfels or andesite hornfels with development of new hornblende and biotite and a partial recrystallization of the original feldspar. Chert or flint may become a finely crystalline quartz rock; sandstones lose their clastic structure and are converted into a mosaic of small close-fitting grains of quartz in a metamorphic rock called quartzite.

If the rock was originally banded or foliated (as, for example, a laminated sandstone or a foliated calc-schist) this character may not be obliterated, and a banded hornfels is the product; fossils even may have their shapes preserved, though entirely recrystallized, and in many contact-altered lavas the vesicles are still visible, though their contents have usually entered into new combinations to form minerals that were not originally present. The minute structures, however, disappear, often completely, if the thermal alteration is very profound. Thus small grains of quartz in a shale are lost or blend with the surrounding particles of clay, and the fine ground-mass of lavas is entirely reconstructed.

By recrystallization in this manner peculiar rocks of very distinct types are often produced. Thus shales may pass into cordierite rocks, or may show large crystals of andalusite (and chiastolite), staurolite, garnet, kyanite and sillimanite, all derived from the aluminous content of the original shale. A considerable amount of mica (both muscovite and biotite) is often simultaneously formed, and the resulting product has a close resemblance to many kinds of schist. Limestones, if pure, are often turned into coarsely crystalline marbles; but if there was an admixture of clay or sand in the original rock such minerals as garnet, epidote, idocrase, wollastonite, will be present. Sandstones when greatly heated may change into coarse quartzites composed of large clear grains of quartz. These more intense stages of alteration are not so commonly seen in igneous rocks, because their minerals, being formed at high temperatures, are not so easily transformed or recrystallized.

In a few cases rocks are fused and in the dark glassy product minute crystals of spinel, sillimanite and cordierite may separate out. Shales are occasionally thus altered by basalt dikes, and feldspathic sandstones may be completely vitrified. Similar changes may be induced in shales by the burning of coal seams or even by an ordinary furnace.

There is also a tendency for metasomatism between the igneous magma and sedimentary country rock, whereby the chemicals in each are exchanged or introduced into the other. Granites may absorb fragments of shale or pieces of basalt. In that case, hybrid rocks called skarn arise, which don't have the characteristics of normal igneous or sedimentary rocks. Sometimes an invading granite magma permeates the rocks around, filling their joints and planes of bedding, etc., with threads of quartz and feldspar. This is very exceptional but instances of it are known and it may take place on a large scale.

## Regional Metamorphism

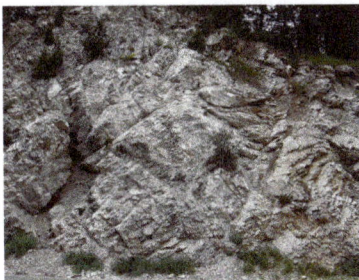

Mississippian marble in Big Cottonwood Canyon, Wasatch Mountains, Utah.

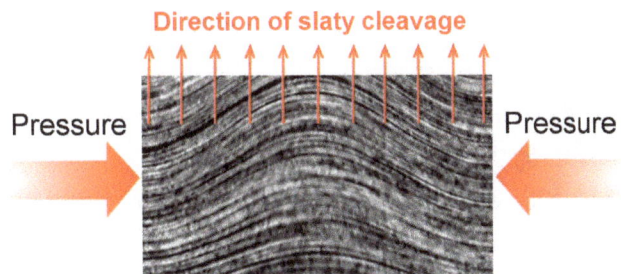

Direction of slaty cleavage

Pressure          Pressure

Dynamic metamorphism

Regional metamorphism tends to make the rock more indurated and at the same time to give it a foliated, shistose or gneissic texture, consisting of a planar arrangement of the minerals, so that platy or prismatic minerals like mica and hornblende have their longest axes arranged parallel to one another. For that reason many of these rocks split readily in one direction along mica-bearing zones (schists). In gneisses, minerals also tend to be segregated into bands; thus there are seams of quartz and of mica in a mica schist, very thin, but consisting essentially of one mineral. Along the mineral layers composed of soft or fissile minerals the rocks will split most readily, and the freshly split specimens will appear to be faced or coated with this mineral; for example, a piece of mica schist looked at facewise might be supposed to consist entirely of shining scales of mica. On the edge of the specimens, however, the white folia of granular quartz will be visible. In gneisses these alternating folia are sometimes thicker and less regular than in schists, but most importantly less micaceous; they may be lenticular, dying out rapidly. Gneisses also, as a rule, contain more feldspar than schists do, and are tougher and less fissile. Contortion or crumbling of the foliation is by no means uncommon; splitting faces are undulose or puckered. Schistosity and gneissic banding (the two main types of foliation) are formed by directed pressure at elevated temperature, and to interstitial movement, or internal flow arranging the mineral particles while they are crystallizing in that directed pressure field.

Rocks that were originally sedimentary and rocks that were undoubtedly igneous may be metamorphosed into schists and gneisses. If originally of similar composition they may be very difficult to distinguish from one another if the metamorphism has been great. A quartz-porphyry, for example, and a fine feldspathic sandstone, may both be metamorphosed into a grey or pink mica-schist.

## Metamorphic Rock Textures

The five basic metamorphic textures with typical rock types are slaty (includes slate and phyllite; the foliation is called "slaty cleavage"), schistose (includes schist; the foliation is called "schistosity"), gneissose (gneiss; the foliation is called "gneissosity"), granoblastic (includes granulite, some marbles and quartzite), and hornfelsic (includes hornfels and skarn).

## Rock Cycle

Earth has witnessed the transformation of the rocks innumerable times in the time span of 4.57 Ga in its history. Earth formed from a ball of melted material during the birth of the solar system. Thereafter the Earth cooled, and solidified at the surface forming a shell of solid igneous rock. While the Earth's crust was forming it witnessed extensive volcanic activity and thus resulted in the formation of igneous rocks. The spewed gases and steam created the first atmosphere and oceans and therefore the first weather. The weather conditions generated on the surface of the planet wore down the igneous rock into sediments. The sediments collected into the low areas or depressions to accumulate as the mountains wore down. In some places the piles of sediment stacked up for hundreds or thousands of feet. The intense weight of the sediments began to coat the sediment grains as mineral cement. The combination of compaction and cementation caused the sediments to transform into solid sedimentary rock in a process called lithification. The sedimentary and igneous rock formed at the bottom was driven deeper into the Earth under increasing pressure and temperature. These rocks underwent recrystallisation and rearrangement thus forming new minerals giving rise to metamorphic rock. Even though a new rock has formed, the process of heating and pressurisation does not stop. Eventually, the minerals reached their melting points and the rocks turned into liquid magma. The return of the melted rock completes the cycle and the rocks go on to become igneous then sedimentary and metamorphic rocks again. We had discussed the basic concepts of rock cycle.

Now let us look at Figure and discuss rock cycle.

Magma occurs as molten material inside the Earth and is the source of all the igneous rocks. Since Earth was largely molten state during its origin, magma may be considered the beginning of the rock cycle. The relationship between igneous, metamorphic and sedimentary rock constitute a "rock cycle" which is a continuous process. Rock cycle is considered to be operating through ages; it is intimately involved with other cyclic Earth processes. This is one of the basic concepts of geology emerging out of the principle of uniformitarianism principle (the present is the key to the past) was given by Hutton in 1785. Rock cycle is particularly closely related to the plate tectonic processes. It starts with cooling and consolidation ofmantle derived magmas at the divergent or convergent boundaries or within intraplate tectonic setting. The erosion of lavas and exposed deep seated rocks produced clastic materials which are transported to low lying depositional basins. The deeply buried sediments in due course are deformed and metamorphosed. The tectonic cycle leads to deformation, reconstitution, uplift and accompanying erosion of fresh rocks so that the cycle continues. The movements of tectonic plates are the main forces driving the rock cycle.

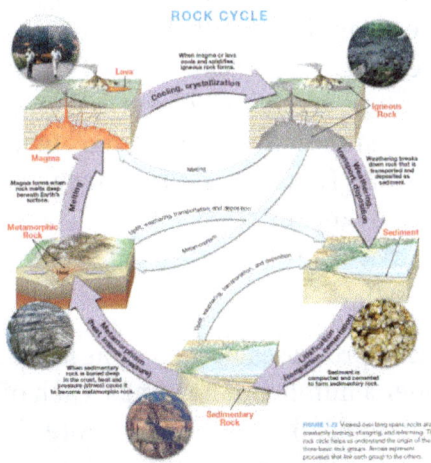

Rock cycle- rocks are constantly forming, changing and reforming.
The rock cycle helps us to understand the origin of three rock groups

The energy changes or redistribution of energy within the Earth systems is manifested by the operation of rock cycle. The rock cycle illustrates the role of various geologic processes operating and transforming from one type into another. The rock cycle helps us to visualise interrelationships among different components of the Earth systems. The rock cycle explains how geological processes can change a rock from one type to another through geological time.

## The Rock Cycle

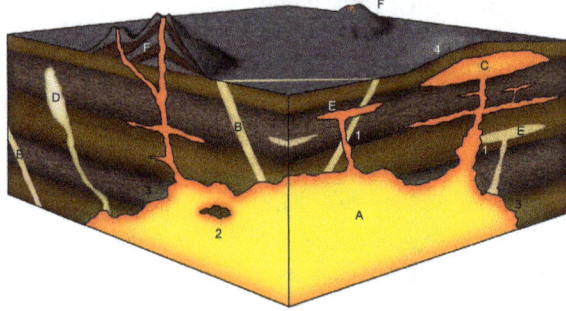

Structures of Igneous Rock. Legend: A = magma chamber (batholith); B = dyke/dike; C = laccolith; D = pegmatite; E = sill; F = stratovolcano; processes: 1 = newer intrusion cutting through older one; 2 = xenolith or roof pendant; 3 = contact metamorphism; 4 = uplift due to laccolith emplacement.

## Transition to Igneous Rock

When rocks are pushed deep under the Earth's surface, they may melt into magma. If the conditions no longer exist for the magma to stay in its liquid state, it cools and solidifies into an igneous rock. A rock that cools within the Earth is called intrusive or plutonic and cools very slowly, producing a coarse-grained texture such as the rock granite. As a result of volcanic activity, magma (which is called lava when it reaches Earth's surface) may cool very rapidly while being on the Earth's surface exposed to the atmosphere and are called extrusive or volcanic rocks. These rocks are fine-grained and sometimes cool so rapidly that no crystals can form and result in a natural glass, such as obsidian, however the most common fine grained rock would be known as basalt. Any of the three main types of rocks (igneous, sedimentary, and metamorphic rocks) can melt into magma and cool into igneous rocks.

## Secondary Changes

Epigenetic change (secondary processes occurring at low temperatures and low pressures) may be arranged under a number of headings, each of which is typical of a group of rocks or rock-forming minerals, though usually more than one of these alterations is in progress in the same rock. Silicification, the replacement of the minerals by crystalline or crypto-crystalline silica, is most common in felsic rocks, such as rhyolite, but is also found in serpentine, etc. Kaolinization is the decomposition of the feldspars, which are the most common minerals in igneous rocks, into kaolin (along with quartz and other clay minerals); it is best shown by granites and syenites. Serpentinization is the alteration of olivine to serpentine (with magnetite); it is typical of peridotites, but occurs in most of the mafic rocks. In uralitization, secondary hornblende replaces augite; chloritization is the alteration of augite (biotite or hornblende) to chlorite, and is seen in many diabases, diorites and greenstones. Epidotization occurs also in rocks of this group, and consists in the development of epidote from biotite, hornblende, augite or plagioclase feldspar.

## Transition to Metamorphic Rock

This diamond is a mineral from within an igneous or metamorphic rock that formed at high temperature and pressure.

Rocks exposed to high temperatures and pressures can be changed physically or chemically to form a different rock, called metamorphic. Regional metamorphism refers to the effects on large masses of rocks over a wide area, typically associated with mountain building events within orogenic belts. These rocks commonly exhibit distinct bands of differing mineralogy and colors, called foliation. Another main type of metamorphism is caused when a body of rock comes into contact with an igneous intrusion that heats up this surrounding country rock. This *contact metamorphism* results in a rock that is altered and re-crystallized by the extreme heat of the magma and/or by the addition of fluids from the magma that add chemicals to the surrounding rock (metasomatism). Any pre-existing type of rock can be modified by the processes of metamorphism.

## Transition to Sedimentary Rock

Rocks exposed to the atmosphere are variably unstable and subject to the processes of weathering and erosion. Weathering and erosion break the original rock down into smaller fragments and carry away dissolved material. This fragmented material accumulates and is buried by additional material. While an individual grain of sand is still a member of the class of rock it was formed from, a rock made up of such grains fused together is sedimentary. Sedimentary rocks can be formed from the lithification of these buried smaller fragments (clastic sedimentary rock), the accumulation and lithification of material generated by living organisms (biogenic sedimentary rock - fossils), or lithification of chemically precipitated material from a mineral bearing solution due to evaporation (precipitate sedimentary rock). Clastic rocks can be formed from fragments broken apart from larger rocks of any type, due to processes such as erosion or from organic material, like plant remains. Biogenic and precipitate rocks form from the deposition of minerals from chemicals dissolved from all other rock types.

## Forces that Drive the Rock Cycle

### Plate Tectonics

In 1967, J. Tuzo Wilson published an article in Nature describing the repeated opening and closing of ocean basins, in particular focusing on the current Atlantic Ocean area. This concept, a part of the plate tectonics revolution, became known as the *Wilson cycle*. The Wilson cycle has had profound effects on the modern interpretation of the rock cycle as plate tectonics became recognized as the driving force for the rock cycle.

### Spreading Ridges

At the mid-ocean divergent boundaries *new* magma is produced by mantle upwelling and a shallow *melting zone*. This *juvenile* basaltic magma is an early phase of the igneous portion of the cycle. As the tectonic plates on either side of the ridge move apart the new rock is carried away from the ridge, the interaction of heated circulating seawater through fractures starts the retrograde metamorphism of the new rock.

### Subduction Zones

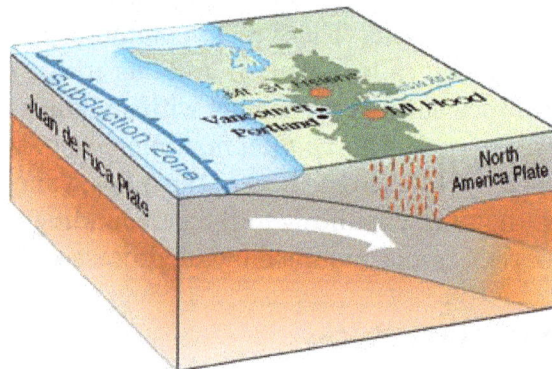

The Juan de Fuca plate sinks below the North America plate at the Cascadia subduction zone.

The new basaltic oceanic crust eventually meets a subduction zone as it moves away from the spreading ridge. As this crust is pulled back into the mantle, the increasing pressure and temperature conditions cause a restructuring of the mineralogy of the rock, this metamorphism alters the rock to form eclogite. As the slab of basaltic crust and some included sediments are dragged deeper, water and other more volatile materials are driven off and rise into the overlying wedge of rock above the subduction zone, which is at a lower pressure. The lower pressure, high temperature, and now volatile rich material in this wedge melts and the resulting buoyant magma rises through the overlying rock to produce island arc or continental margin volcanism. This volcanism includes more silicic lavas the further from the edge of the island arc or continental margin, indicating a deeper source and a more differentiated magma.

At times some of the metamorphosed downgoing slab may be thrust up or obducted onto the continental margin. These blocks of mantle peridotite and the metamorphic eclogites are exposed as ophiolite complexes.

The newly erupted volcanic material is subject to rapid erosion depending on the climate conditions. These sediments accumulate within the basins on either side of an island arc. As the sediments become more deeply buried lithification begins and sedimentary rock results.

## Continental Collision

On the closing phase of the classic Wilson cycle, two continental or smaller terranes meet at a convergent zone. As the two masses of continental crust meet, neither can be subducted as they are both *low density* silicic rock. As the two masses meet, tremendous compressional forces distort and modify the rocks involved. The result is regional metamorphism within the interior of the ensuing orogeny or mountain building event. As the two masses are compressed, folded and faulted into a mountain range by the continental collision the whole suite of pre-existing igneous, volcanic, sedimentary and earlier metamorphic rock units are subjected to this new metamorphic event.

## Accelerated Erosion

The high mountain ranges produced by continental collisions are immediately subjected to the forces of erosion. Erosion wears down the mountains and massive piles of sediment are developed in adjacent ocean margins, shallow seas, and as continental deposits. As these sediment piles are buried deeper they become lithified into sedimentary rock. The metamorphic, igneous, and sedimentary rocks of the mountains become the new piles of sediments in the adjoining basins and eventually become sedimentary rock.

## An Evolving Process

The plate tectonics rock cycle is an evolutionary process. Magma generation, both in the spreading ridge environment and within the wedge above a subduction zone, favors the eruption of the more silicic and volatile rich fraction of the crustal or upper mantle material. This lower density material tends to stay within the crust and not be subducted back into the mantle. The magmatic aspects of plate tectonics tends to gradual segregation within or between the mantle and crust. As magma forms, the initial melt is composed of the more silicic phases that have a lower melting point. This leads to partial melting and further segregation of the lithosphere. In addition the silicic continental crust is relatively buoyant and is not normally subducted back into the mantle. So over time the continental masses grow larger and larger.

## The Role of Water

The presence of abundant water on Earth is of great importance for the rock cycle. Most obvious perhaps are the water driven processes of weathering and erosion. Water in the form of precipitation and acidic soil water and groundwater is quite effective at dissolving minerals and rocks, especially those igneous and metamorphic rocks and marine sedimentary rocks that are unstable under near surface and atmospheric conditions. The water carries away the ions dissolved in solution and the broken down fragments that are the products of weathering. Running water carries vast amounts of sediment in rivers back to the ocean and inland basins. The accumulated and buried sediments are converted back into rock.

A less obvious role of water is in the metamorphism processes that occur in fresh seafloor volcanic rocks as seawater, sometimes heated, flows through the fractures and crevices in the rock. All of these processes, illustrated by serpentinization, are an important part of the destruction of volcanic rock.

The role of water and other volatiles in the melting of existing crustal rock in the wedge above a subduction zone is a most important part of the cycle. Along with water, the presence of carbon dioxide and other carbon compounds from abundant marine limestone within the sediments atop the down going slab is another source of melt inducing volatiles. This involves the carbon cycle as a part of the overall rock cycle.

## References

- Dietrich, Richard Vincent; Skinner, Brian J. (2009), Gems, Granites, and Gravels: knowing and using rocks and minerals (Cambridge University Press) ISBN 978-0-521-10722-8

- Roberts, Dar. "Rocks and classifications". Department of Geography, University of California, Santa Barbara. Archived from the original on 31 October 2012. Retrieved 11 November 2012

- Philpotts, Anthony; Ague, Jay (2009), Principles of Igneous and Metamorphic Petrology (Cambridge University Press) ISBN 978-0-521-88006-0

- Oxford Academic: Crustal Contamination of Picritic Magmas During Transport Through Dikes: the Expo Intrusive Suite, Cape Smith Fold Belt, New Quebec | Journal of Petrology | Oxford Academic, accessdate: March 27, 2017

- Terrascope. "Environmental Risks of Mining". The Future of strategic Natural Resources. Cambridge, Massachusetts, USA: Massachusetts Institute of Technology. Archived from the original on 20 September 2014. Retrieved 10 September 2014

- Prothero, Donald R.; Schwab, Fred (2004). Sedimentary geology : an introduction to sedimentary rocks and stratigraphy (2nd ed.). New York: Freeman. p. 12. ISBN 978-0-7167-3905-0

# 3

# An Introduction to Structural Geology

Structural geology is the study of the measurements of rock geometries in order to gain insights into their deformation histories, and understand the stress fields. Various concepts and methods, which have been thoroughly explained in this chapter, include outcrop, layering, intrusion, deformation, strike and dip, and rock structures, to provide an extensive understanding of the subject.

Structural geology is the branch of geology that deals with the recognition, representation, and genetic interpretation of rock structures. It also involves the study of the forces which give rise to these structures. The term 'Structural' is derived from Latin word '*Struere*' which means 'to build'.

Structural geology is the study of the architecture of rocks insofar as it has resulted from deformation (Billings, 1990).

Structural geology deals with the geometry, distribution and formation of structures (Fossen, 2010).

Geologic structures or rock structures incorporate symmetry and geometric configuration of rocks present in the Earth's crust on all scales. Geologic structures result from the deformation caused by the tectonic forces present in Earth, i.e., they are endogenic. The term tectonics is derived from Greek word '*Tektos*' meaning 'builder'. Tectonics is the study of the forces and motion that result in rock deformation and structure.

## Relationship between Structural Geology and Tectonics

Let us get acquainted with the relation between structural geology and tectonics. Structural geology is mainly concerned with the rock geometry whereas, tectonics deals with the forces and movements responsible for the generation of the rock or geologic structures. Therefore both structural geology and tectonics are responsible for building up the Earth's lithosphere. We can say that tectonics is quite closely connected to the underlying processes that cause geologic structures to form. These structures provide information about the forces acting within the Earth. The objective of structural geology is to determine and explain the architecture of rocks as observed in the field. The field observations are supported by laboratory investigations to attain this objective. Geologic structures range in size from microscopic scale - to hundreds of kilometres.

## Use and Importance

The study of geologic structures has been of prime importance in economic geology, both petroleum geology and mining geology. Folded and faulted rock strata commonly form traps that accumulate and concentrate fluids such as petroleum and natural gas. Similarly, faulted and structurally complex areas are notable as permeable zones for hydrothermal fluids, resulting in concentrated areas of base and precious metal ore deposits. Veins of minerals containing various metals commonly occupy faults and fractures in structurally complex areas. These structurally fractured and faulted zones often occur in association with intrusive igneous rocks. They often also occur around geologic reef complexes and collapse features such as ancient sinkholes. Deposits of gold, silver, copper, lead, zinc, and other metals, are commonly located in structurally complex areas.

Structural geology is a critical part of engineering geology, which is concerned with the physical and mechanical properties of natural rocks. Structural fabrics and defects such as faults, folds, foliations and joints are internal weaknesses of rocks which may affect the stability of human engineered structures such as dams, road cuts, open pit mines and underground mines or road tunnels.

Geotechnical risk, including earthquake risk can only be investigated by inspecting a combination of structural geology and geomorphology. In addition, areas of karst landscapes which reside atop underground caverns, potential sinkholes, or other collapse features are of particular importance for these scientists. In addition, areas of steep slopes are potential collapse or landslide hazards.

Environmental geologists and hydrogeologists need to apply the tenets of structural geology to understand how geologic sites impact (or are impacted by) groundwater flow and penetration. For instance, a hydrogeologist may need to determine if seepage of toxic substances from waste dumps is occurring in a residential area or if salty water is seeping into an aquifer.

Plate tectonics is a theory developed during the 1960s which describes the movement of continents by way of the separation and collision of crustal plates. It is in a sense structural geology on a planet scale, and is used throughout structural geology as a framework to analyze and understand global, regional, and local scale features.

# Concepts in Structural Geology

Let us get familiarised with some of the basic concepts of structural geology before we study about various geologic or rock structures, like folds, faults, *etc.*

The study of structural geology is mainly accomplished with the study of geological structures which are developed due to deformation. They are also known as deformational structures or secondary structures. Primary structures form as the result of the processes connected with the deposition of sediments. They are also known as depositional structures, such as bedding plane.

Structural geologists play a significant role in the identification of the geologic structures, their geometry and orientation, the time and sequence of their deposition. They also assess the physical conditions responsible for the development of these structures.

Let us read about the common terminologies used in structural geology.

## Outcrop

They are exposure of rocks on the surface of Earth (Figure). If the rocks are not exposed on the surface of Earth, they are referred to as 'incrop'. The outcrops are generally visible along valley walls, river/*nala* section, mountains, road/ railway cuttings and tunnels. Hence an outcrop denotes area on surface of the Earth, over which a rock mass is exposed. The line of intersection of the limiting surface of rock mass with surface of ground marks its' limit. The study of geological structures is generally carried by the study of outcrops.

Outcrop of sandstone, note the horizontal bedding planes

An outcrop or rocky outcrop is a visible exposure of bedrock or ancient superficial deposits on the surface of the Earth.

Granite outcrops at Silesian Stones Mountain in southwestern Poland.

View of a bedrock outcrop near San Carlos Water, Falkland Islands

Serrote Branco's outcrop. Caicó, Brazil.

## Features

Outcrops do not cover the majority of the Earth's land surface because in most places the bedrock or superficial deposits are covered by a mantle of soil and vegetation and cannot be seen or examined closely. However, in places where the overlying cover is removed through erosion or tectonic uplift, the rock may be exposed, or *crop out*. Such exposure will happen most frequently in areas where erosion is rapid and exceeds the weathering rate such as on steep hillsides, mountain ridges and tops, river banks, and tectonically active areas. In Finland, glacial erosion during the last glacial maximum (ca. 11000 BC), followed by scouring by sea waves, followed by isostatic uplift has produced a large number of smooth coastal and littoral outcrops.

A typical shore outcrop scoured by ancient glaciers in Espoo, Finland.

Bedrock and superficial deposits may also be exposed at the Earth's surface due to human excavations such as quarrying and building of transport routes.

## Study

Outcrops allow direct observation and sampling of the bedrock *in situ* for geologic analysis and creating geologic maps. In situ measurements are critical for proper analysis of geological history and outcrops are therefore extremely important for understanding the geologic time scale of earth history. Some of the types of information that cannot be obtained except from bedrock outcrops or by precise drilling and coring operations, are structural geology features orientations (e.g. bedding planes, fold axes, foliation), depositional features orientations (e.g. paleo-current directions, grading, facies changes), paleomagnetic orientations. Outcrops are also very important for understanding fossil assemblages, and paleo-environment, and evolution as they provide a record of relative changes within geologic strata.

Accurate description, mapping, and sampling for laboratory analysis of outcrops made possible all of the geologic sciences and the development of fundamental geologic laws such as the law of superposition, the principle of original horizontality, principle of lateral continuity, and the principle of faunal succession.

## Examples

On Ordnance Survey maps in Great Britain, cliffs are distinguished from outcrops: cliffs have a continuous line along the top edge with lines protruding down; outcrops have a continuous line around each area of bare rock. An outcrop example in California is the Vasquez Rocks, familiar from location shooting use in many films, composed of uplifted sandstone. Yana is another example of outcrops, located in Uttara Kannada district in Karnataka, India.

## Layering

It is developed in rocks due to the deposition of sediments, rock materials or minerals. The deposition of sediments one over another in a basin results in layered or sedimentary rocks. Bed is defined as an individual layer or strata of a sedimentary rock (Figure). Each bed is separated by the adjacent bed by a plane called bedding plane. If the individual layers are less than one cm thick they are called lamination. The process of deposition of sediments in a layer by layer fashion is known as stratification.

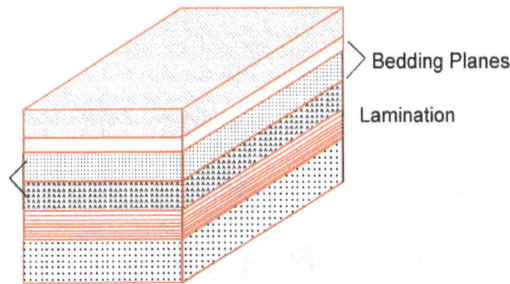

Stratification with numerous beds, bedding planes and lamination

The bedding planes are found in sedimentary rocks. Layering is developed in volcanic rocks (igneous rocks) (Figure) because of flow of lava. It is also referred as primary foliation because these are developed simultaneously with formation of the rock. The term secondary foliation is used for layers found in metamorphic rocks (Figure). The layers in metamorphic rocks develop as a result of the development of new minerals and reorientation of mineral particles from pre-existing rocks during the process of metamorphism. Slaty cleavage, schistosity, gneissosity, *etc.* fall in the category of secondary foliation.

(a)                                                                                          (b)

(a) Primary foliation developed in igneous rocks, and (b) Secondary foliation developed in metamorphic rocks

## Stratum

In geology and related fields, a stratum (plural: strata) is a layer of sedimentary rock or soil, or igneous rock that were formed at the Earth's surface, with internally consistent characteristics that distinguish it from other layers. The "stratum" is the fundamental unit in a stratigraphic column and forms the basis of the study of stratigraphy.

Strata in Salta (Argentina).

Goldenville strata in quarry in Bedford, Canada. These are Middle Cambrian marine sediments. This formation covers over half of Nova Scotia and is recorded as being 29,000 feet thick in some areas.

## Characteristics

Each layer is generally one of a number of parallel layers that lie one upon another, laid down by natural processes. They may extend over hundreds of thousands of square kilometers of the Earth's surface. Strata are typically seen as bands of different colored or differently structured material exposed in cliffs, road cuts, quarries, and river banks. Individual bands may vary in thickness from a few millimeters to a kilometer or more. Each band represents a specific mode of deposition: river silt, beach sand, coal swamp, sand dune, lava bed, etc.

The Permian through Jurassic strata in the Colorado Plateau area of southeastern Utah demonstrates the principles of stratigraphy.

These strata make up much of the famous prominent rock formations in widely spaced protected areas such as Capitol Reef National Park and Canyonlands National Park.

## Naming

Geologists study rock strata and categorize them by the material of beds. Each distinct layer is typically assigned to the name of sheet, usually based on a town, river, mountain, or region where the formation is exposed and available for study. For example, the Burgess Shale is a thick exposure of dark, occasionally fossiliferous, shale exposed high in the Canadian Rockies near Burgess Pass. Slight distinctions in material in a formation may be described as "members" (or sometimes "beds"). Formations are collected into "groups" while groups may be collected into "supergroups".

## Beds

Beds are the layers of sedimentary rocks that are distinctly different from overlying and underlying subsequent beds of different sedimentary rocks. Layers of beds are called stratigraphy or strata. They are formed from sedimentary rocks being deposited on the Earth's solid surface over a long periods of time. The stratigraphy are layered in the same order that they were deposited, allowing a differentiation of which beds are younger and which ones are older (the Law of Superposition). The structure of a bed is determined by its bedding plane. Beds can be differentiated in various ways, including rock or mineral type and particle size. The term is generally applied to sedimentary strata, but may also be used for volcanic flows or ash layers.

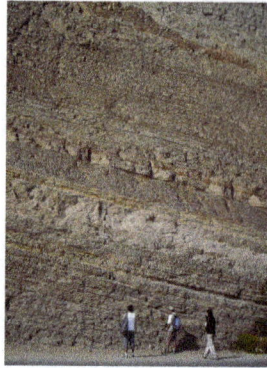

Tilted sedimentary bedding in shales of the Cretaceous Salto del Fraile Formation, Peru.

In a quarry, a bedding is a term used for a structure occurring in granite and similar massive rocks that allows them to split in well-defined planes horizontally or parallel to the land surface. Other kinds of beds are cross beds and graded beds. Cross beds are not layered horizontally and are formed by a combination of local deposition on the inclined surfaces of ripples or dunes, and local erosion. Graded beds shows a gradual change in grain or clast sizes from one side of the bed to the other. A normal grading is when there are bigger grain sizes on the older side while an inverse grading is when there are smaller grain sizes on the older side. By knowing the type of beds, geologists can determine the relative ages of the rocks.

## Bed Thickness

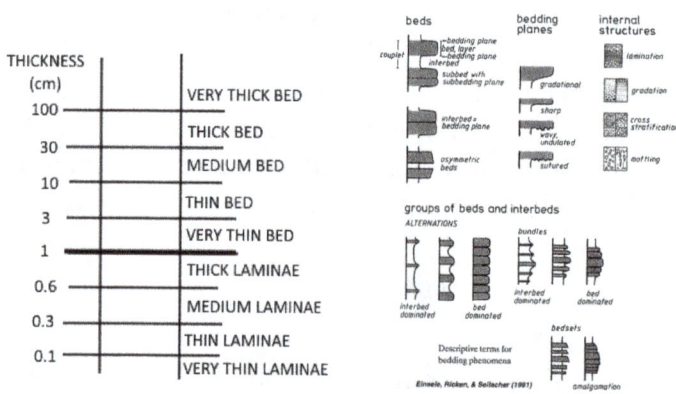

Thickness of bed and laminae sizes in centimeters

A bed is the smallest lithostratigraphic unit, usually ranging in thickness from a centimeter to several meters and distinguishable from beds above and below it. The thickness of the bed is determined by the time period involving the deposition of the rocks.

- Very Thick Bed - 100cm

- Thick Bed - 30cm

- Medium Bed - 10cm

- Thin Bed - 3cm

- Very Thin Bed - 1cm

- Thinner than 1cm is called a Lamina

## Geologic Principles

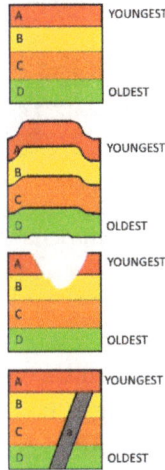

Law of Superposition, Law of Original Horizontality, Law of Lateral Continuity,
Cross-Cutting Relationship

There are geologic principles that the beds normally follow. Even though there can be cases where the principles do not apply mostly due to faults, they are true for most cases.

- Law of Superposition is the law that states that the oldest rocks are deposited first and has the younger layers deposited last, as long as the beds have not been overturned through tectonic activities. This is used to date the stratigraphy and their relative ages.

- Law of Original Horizontality states that if the beds are not horizontal, then the layers were caused to either fold or tilt through tectonic activities. They were all deposited horizontally due to gravity.

- Law of Lateral Continuity states that the bed deposits extends in all lateral directions. This means that if two places separated by erosional features have similar rocks, it could mean that they were originally continuous.

- Cross-Cutting Relationship states that a fault is younger than the rock layers that it goes through. It helps with relatively dating the rocks.

# Lamination

In geology, lamination is a small-scale sequence of fine layers (*laminae*; singular: *lamina*) that occurs in sedimentary rocks. Laminae are normally smaller and less pronounced than bedding. Lamination is often regarded as planar structures one centimetre or less in thickness, whereas bedding layers are greater than one centimetre. However, structures from several millimetres to many centimetres have been described as laminae. A single sedimentary rock can have both laminae and beds.

Lamination in a piece of travertine. In this case the layering was caused by seasonal differences in sediment supply. This rock was part of the Roman aqueduct of Mons/Montauroux - Fréjus and was most probably derived from the karst area in the vicinity.

## Description

Lamination consists of small differences in the type of sediment that occur throughout the rock. They are caused by cyclic changes in the supply of sediment. These changes can occur in grain size, clay percentage, microfossil content, organic material content or mineral content and often result in pronounced differences in colour between the laminae. Weathering can make the differences even more clear.

Lamination can occur as parallel structures (parallel lamination) or in different sets that make an angle with each other (cross-lamination). It can occur in many different types of sedimentary rock, from coarse sandstone to fine shales, mudstones or in evaporites.

Because lamination is a small structure, it is easily destroyed by bioturbation (the activity of burrowing organisms) shortly after deposition. Lamination therefore survives better under anoxic circumstances, or when the sedimentation rate was high and the sediment was buried before bioturbation could occur.

## Origin

Lamination develops in fine grained sediment when fine grained particles settle, which can only happen in quiet water. Examples of sedimentary environments are deep marine (at the seafloor) or lacustrine (at the bottom of a lake), or mudflats, where the tide creates cyclic differences in sediment supply.

Laminae formed in glaciolacustrine environments (in glacier lakes) are a special case. They are called varves. Quaternary varves are used in stratigraphy and palaeoclimatology to reconstruct climate changes during the last few hundred thousand years.

Lamination in sandstone is often formed in a coastal environment, where wave energy causes a separation between grains of different sizes.

## Foliation

Gneiss, a foliated metamorphic rock.

Foliation in geology refers to repetitive layering in metamorphic rocks. Each layer can be as thin as a sheet of paper, or over a meter in thickness. The word comes from the Latin *folium*, meaning "leaf", and refers to the sheet-like planar structure. It is caused by shearing forces (pressures pushing different sections of the rock in different directions), or differential pressure (higher pressure from one direction than in others). The layers form parallel to the direction of the shear, or perpendicular to the direction of higher pressure. Nonfoliated metamorphic rocks are typically formed in the absence of significant differential pressure or shear. Foliation is common in rocks affected by the regional metamorphic compression typical of areas of mountain belt formation (orogenic belts).

More technically, foliation is any penetrative planar fabric present in metamorphic rocks. Rocks exhibiting foliation include the standard sequence formed by the prograde metamorphism of mudrocks; slate, phyllite, schist and gneiss. The *slatey cleavage* typical of slate is due to the preferred orientation of microscopic phyllosilicate crystals. In gneiss, the foliation is more typically represented by compositional banding due to segregation of mineral phases. Foliated rock is also known as S-tectonite in sheared rock masses.

Examples include the bands in gneiss (gneissic banding), a preferred orientation of planar *large* mica flakes in schist (Schistocity), the preferred orientation of *small* mica flakes in phyllite (with its planes having a silky sheen, called *phylitic luster* – the Greek word, *phyllon*, also means "leaf"), the extremely fine grained preferred orientation of clay flakes in slate (called "slaty cleavage"), and the layers of flattened, smeared, pancake-like clasts in metaconglomerate.

## Formation Mechanisms

Foliation is usually formed by the preferred orientation of minerals within a rock.

Usually this is a result of some physical force, and its effect upon the growth of minerals. The planar fabric of a foliation typically forms at right angles to the maximum principal strain direction. In sheared zones, however, planar fabric within a rock may not be directly perpendicular to the principal stress direction due to rotation, mass transport and shortening.

Foliation may be formed by realignment of micas and clays via physical rotation of the minerals within the rock. Often this foliation is associated with diagenetic metamorphism and low-grade burial metamorphism. Foliation may parallel original sedimentary bedding, but more often is oriented at some angle to it.

The growth of platy minerals, typically of the mica group, is usually a result of prograde metamorphic reactions during deformation. Often, retrograde metamorphism will not form a foliation because unroofing of a metamorphic belt is not accompanied by significant compressive stress. Thermal metamorphism in the aureole of a granite is also unlikely to result in growth of mica in a foliation, although growth of new minerals may overprint existing foliation(s).

Alignment of tabular minerals in metamorphic rocks, igneous rocks and intrusive rocks may form a foliation. Typical examples of metamorphic rocks include porphyroblastic schists where large, oblate minerals form an alignment either due to growth or rotation in the groundmass.

Igneous rocks can become foliated by alignment of cumulate crystals during convection in large magma chambers, especially ultramafic intrusions, and typically plagioclase laths. Granite may form foliation due to frictional drag on viscous magma by the wall rocks. Lavas may preserve a flow foliation, or even compressed eutaxitic texture, typically in highly viscous felsic agglomerate, welded tuff and pyroclastic surge deposits.

Metamorphic differentiation, typical of gneisses, is caused by chemical and compositional banding within the metamorphic rock mass. Usually this represents the protolith chemistry, which forms distinct mineral assemblages. However, compositional banding can be the result of nucleation processes which cause chemical and mineralogical differentiation into bands. This typically follows the same principle as mica growth, perpendicular to the principal stress. Metamorphic differentiation can be present at angles to protolith compositional banding.

*Crenulation cleavage* and *oblique foliation* are particular types of foliation.

### Interpretation

Foliation, as it forms generally perpendicular to the direction of principal stress, records the direction of shortening. This is related to the axis of folds, which generally form an *axial-planar* foliation within their axial regions.

Measurement of the intersection between a fold's axial plane and a surface on the fold will provide the fold plunge. If a foliation does not match the observed plunge of a fold, it is likely associated with a different deformation event.

Foliation in areas of shearing, and within the plane of thrust faults, can provide information on the transport direction or sense of movement on the thrust or shear. Generally, the acute intersection angle shows the direction of transport. Foliations typically bend or curve into a shear, which provides the same information, if it is of a scale which can be observed.

Foliations, in a regional sense, will tend to curve around rigid, incompressible bodies such as granite. Thus, they are not always 'planar' in the strictest sense and may violate the rule of being perpendicular to the regional stress field, due to local influences. This is a megascopic version of what may occur around porphyroblasts. Often, fine observation of foliations on outcrop, hand specimen and on the microscopic scale complements observations on a map or regional scale.

## Description

When describing a foliation it is useful to note

- the mineralogy of the folia; this can provide information on the conditions of formation
- the mineralogy in intrafolial areas
- foliation spacing
- any porphyroblasts or minerals associated with the foliation and whether they overprint it or are cut by it
- whether it is planar, undulose, vague or well developed
- its orientation in space, as strike and dip, or dip and dip direction
- its relationship to other foliations, to bedding and any folding
- measure intersection lineations

Following such a methodology allows eventual correlations in style, metamorphic grade, and intensity throughout a region, relationship to faults, shears, structures and mineral assemblages.

## Engineering Considerations

In geotechnical engineering a foliation plane may form a discontinuity that may have a large influence on the mechanical behavior (strength, deformation, etc.) of rock masses in, for example, tunnel, foundation, or slope construction.

## Intrusion

Igneous bodies can be extrusive or intrusive. The intrusive rocks are those igneous rocks which penetrate or intrude in a pre-existing host rock. They might have consolidated either at great depth or at shallow depth. Many times the layering seen in the intrusive igneous rocks which can be observed in parallelism with the surrounding country rocks (also called host rock). But sometimes igneous bodies cut across the beds of the country rock and because of their late entry into the system along some joints or fractures. We will discuss various types of intrusions based on their geometry, later in this module.

## Types of Intrusions

## Dykes

Dykes are sheet-like bodies of igneous rock that cut across sedimentary bedding or foliations in rocks. They may be single or multiple in nature. Dykes often have a fine-grained margin where they were chilled rapidly against the wallrocks. Small dykes tend to be fine-grained while large dykes tend to be coarser grained.

## Silica-rich Intrusions

- Stoped stocks

Stoping happens when a rising magma breaks off jointed blocks from the overlying country rock. The magma forces its way into the cracked roof and fragments of the wallrock sink into the magma. These fragments within the magmatic rock are known as xenoliths and can range in size from less than a millimetre up to tens of kilometres. The xenolith-rich Corralillo granodiorite of Peru is a good example, where xenoliths are concentrated near the roof of the intrusions (roof pendants) but become smaller and more rounded when digested further within the intrusion.

In the Thorr granodioritic pluton of Donegal, Ireland many of the xenoliths are large (several hundred metres) and their origin (e.g. limestone, schist) is still recognisable and can be correlated with the surrounding rock structures. This means there was little movement in the xenoliths after they were incorporated from their parent rocks.

- Ring Dykes and Bell-jar Plutons

Ring dykes form a cylinder around a subsided block of country rock, fill ing the ring-shaped fracture with magma. Excellent examples of ring dykes are seen on the island of Mull, Scotland. Large subsidence of a core of country rock can also form a circular pluton in the shape of a bell-jar.

- Centred Complexes

In composite intrusions, some are arranged in concentric rings. These often formed as rings of intrusion spread outwards from a focus. These are called centred complexes.

- Sheeted Intrusions

Some intrusions form relatively flat-lying sheets, often with an undulating surface.

- Diapiric plutons

Diapirs are circular-shaped intrusions with vertical walls that force their way into place, strongly deforming the surrounding country rocks. The diapirs probably rose as buoyant magma, forcing their way up and through the denser country rocks. Diapirs are generally unfoliated (unbanded) in their centre. They become more foliated (banded) towards the contact with the country rocks, where numerous xenoliths have become more flattened. The foliation and flattening of the xenoliths runs parallel to this contact. The adjacent surrounding country rocks are intensely deformed and stretched out parallel to the intrusion.

- Batholiths

In many parts of the world, granites extend for many hundreds of kilometres in large masses called batholiths. These are not single intrusions but usually composite intrusions of similar magmas. The individual plutons in batholiths vary in size but the largest are about 30 km across. The individual plutons can show quite different forms and ages of emplacement (sometimes separated by millions of years). At the surface, granites in batholiths often weather and erode into rounded masses called tors.

The massive batholiths represent repeated and voluminous production of magmas during a period of plate tectonic activity. An excellent example is the Berridale Batholith of the Snowy Mountains, New South Wales.

## Silica-poor Intrusions

These denser magmas crystallise into several different shapes:

- Flat-lying Sheets

Flat-lying sheets (sills) range in size from less than a metre thick up to huge intrusions underlying thousands of square kilometres. The Whin Sill of northern England is up to 100 m thick and intrudes an area of over 5000 square kilometres. Although sills typically intrude along the surrounding rock layers, almost all large sills vary in thickness and transgress into higher layers when mapped over a large area. These transgressions can occur as abrupt steps. The dolerite sills of the Karoo South Africa, the Transantarctic mountains and Tasmania occur as undulating discordant sheets.

- Laccoliths

Laccoliths are lens-shaped intrusions where magmas were emplaced like a sill between sedimentary layers but then bulged up into a dome. This commonly happens in dioritic intrusions. An excellent example of a laccolith is the Prospect intrusion of Sydney, New South Wales.

- Cone Sheets

Cones are thin intrusive sheets that expand upwards and outwards in cones. Individual sheets are generally a few metres thick, but arranged so that the outer cones dip at lower angles to the inner ones so that they all converge towards a common source at depth.

- Funnel-shaped Intrusions

These have a much thicker top that plunges down to a very narrow neck. A classic example is the Skaergaard intrusion of Greenland.

- Funnel Dykes

Funnel dykes are elongated in outcrop like a dyke but with a V-shaped cross-section that narrows downwards. All known examples are very large (over 100 km in length). The Great Dyke of Zimbabwe is over 700 km long and the Jimberlana dyke of Western Australia is over 180 km long.

- Lopoliths

Lopoliths are the largest known intrusions of dense magma and form a thick saucer shape within the surrounding country rocks. A famous example is the Bushveld Complex of South Africa, which is over 550 km across and up to 8 km in thickness. Lopoliths contain many important economic deposits of nickel, copper, platinum, palladium and chromium. The Sudbury intrusion of Ontario, Canada formed in an oval-shaped depression probably caused by a large meteorite impact.

- Dyke Swarms

Dykes reach their highest numbers in dyke swarms, which may form lines or radial patterns. Radial dykes usually converge on volcanic centres or igneous intrusions. Linear dyke swarms are more extensive than radial dyke swarms but can also be concentrated around large intrusions or volcanic centres. In south-eastern Iceland, the Cenozoic lava succession is cut by thousands of aligned near-vertical basaltic dykes averaging less than a metre thick.

- Diatremes

Diatremes are steep pipe-like bodies filled with fragments of both igneous rocks and wallrocks. They form by explosion which results from the release of contained carbon dioxide gas ($CO_2$) and water vapour ($H_2O$) near the surface. Some very silica-poor magma types are economically important as the host rocks for diamonds.

# Deformation

This term applied for a process in which the original rock is modified. The original rock body is usually undeformed and can be modified by folding, jointing, faulting or under the effect of gravity. Let us discuss the three types of deformation that are caused mainly by the horizontal movements of the lithospheric plates relative to one another. These forces are:

- Tensional forces stretch and pull rock masses apart,

- Compressive forces squeeze and shorten rock masses,

- Shearing forces push two parts of rock masses in opposite directions.

If a rock mass is subjected to directed force for a short period of time, it usually undergoes through three stages of deformation:

- Initially the deformation is elastic and on withdrawal of stress the body returns to its original form. The limiting stress called elastic limit.

- The deformation is plastic when the stress exceeds the elastic limit; that is the body only partially returns to its original shape even if the stress is withdrawn.

- When there is a stress continuously increases, one or more fracture develops and the body eventually fails by rupture.

Ductile and brittle are two commonly used terms by the structural geologist. Brittle deformation relates to the fracturing of the rock. Ductile deformation indicates bending, stretching, folding, thinning of rocks, and realigning of grains.

## Brittle-Ductile Properties of the Lithosphere

We all know that rocks near the surface of the Earth behave in a brittle manner. Crustal rocks are composed of minerals like quartz and feldspar which have high strength, particularly at low pressure and temperature. As we go deeper in the Earth the strength of these rocks initially increases.

At a depth of about 15 km we reach a point called the brittle-ductile transition zone. Below this point rock strength decreases because fractures become closed and the temperature is higher, making the rocks behave in a ductile manner. At the base of the crust the rock type changes to peridotite which is rich in olivine. Olivine is stronger than the minerals that make up most crustal rocks, so the upper part of the mantle is again strong. But, just as in the crust, increasing temperature eventually predominates and at a depth of about 40 km the brittle-ductile transition zone in the mantle occurs. Below this point rocks behave in an increasingly ductile manner.

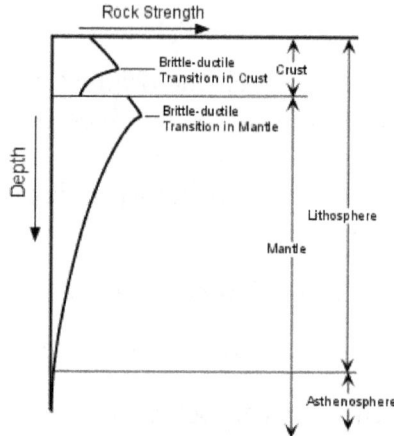

Deformation in Progress

Only in a few cases does deformation of rocks occur at a rate that is observable on human time scales. Abrupt deformation along faults, usually associated with earthquakes occurs on a time scale of minutes or seconds. Gradual deformation along faults or in areas of uplift or subsidence can be measured over periods of months to years with sensitive measuring instruments.

## Evidence of Past Deformation

Evidence of deformation that has occurred in the past is very evident in crustal rocks. For example, sedimentary strata and lava flows generally follow the law of original horizontality. Thus, when we see such strata inclined instead of horizontal, evidence of an episode of deformation.

Since many geologic features are planar in nature, we a way to uniquely define the orientation of a planar feature we first need to define two terms - strike and dip.

For an inclined plane the *strike* is the compass direction of any horizontal line on the plane. The *dip* is the angle between a horizontal plane and the inclined plane, measured perpendicular to the direction of strike.

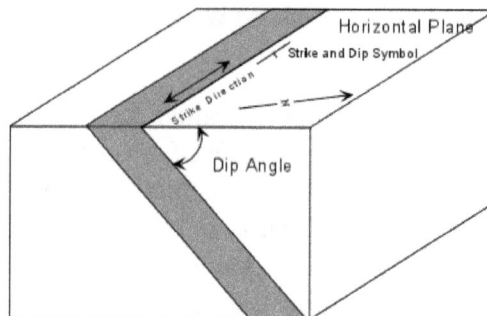

In recording strike and dip measurements on a geologic map, a symbol is used that has a long line oriented parallel to the compass direction of the strike. A short tick mark is placed in the center of the line on the side to which the inclined plane dips, and the angle of dip is recorded next to the strike and dip symbol as shown above. For beds with a 90° dip (vertical) the short line crosses the strike line, and for beds with no dip (horizontal) a circle with a cross inside is used as shown below.

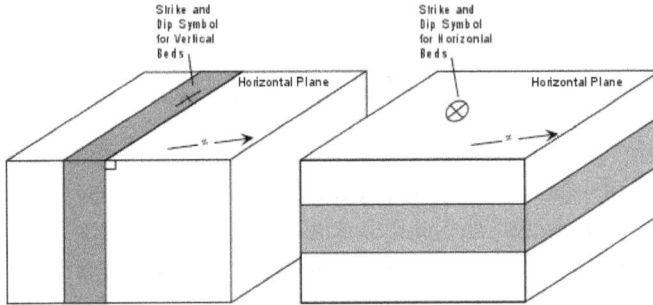

For linear structures, a similar method is used, the strike or bearing is the compass direction and angle the line makes with a horizontal surface is called the plunge angle.

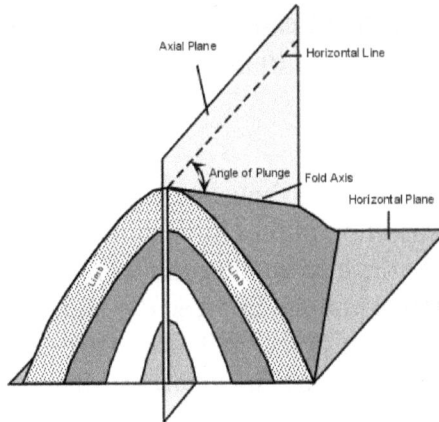

Fracture of Brittle Rocks

As we have discussed previously, brittle rocks tend to fracture when placed under a high enough stress. Such fracturing, while it does produce irregular cracks in the rock, sometimes produces planar features that provide evidence of the stresses acting at the time of formation of the cracks. Two major types of more or less planar fractures can occur: joints and faults.

## Joints

As we learned in our discussion of physical weathering, joints are fractures in rock that show no slippage or offset along the fracture. Joints are usually planar features, so their orientation can be described as a strike and dip. They form from as a result of extensional stress acting on brittle rock. Such stresses can be induced by cooling of rock (volume decreases as temperature decreases) or by relief of pressure as rock is eroded above thus removing weight.

Joints provide pathways for water and thus pathways for chemical weathering attack on rocks. If new minerals are precipitated from water flowing in the joints, this will form a vein. Many veins observed in rock are mostly either quartz or calcite, but can contain rare minerals like gold and silver. These aspects will be discussed in more detail when we talk about valuable minerals from the earth in a couple of weeks.

Because joints provide access of water to rock, rates of weathering and/or erosion are usually higher along joints and this can lead to differential erosion.

From an engineering point of view, joints are important structures to understand. Since they are zones of weakness, their presence is critical when building anything from dams to highways. For dams, the water could leak out through the joints leading to dam failure. For highways the joints may separate and cause rock falls and landslides.

## Faults

Faults occur when brittle rocks fracture and there is an offset along the fracture. When the offset is small, the displacement can be easily measured, but sometimes the displacement is so large that it is difficult to measure.

## Types of Faults

As we found out in our discussion of earthquakes, faults can be divided into several different types depending on the direction of relative displacement. Since faults are planar features, the concept of strike and dip also applies, and thus the strike and dip of a fault plane can be measured. One division of faults is between dip-slip faults, where the displacement is measured along the dip direction of the fault, and strike-slip faults where the displacement is horizontal, parallel to the strike of the fault. Recall the following types of faults:

- Dip Slip Faults - Dip slip faults are faults that have an inclined fault plane and along which the relative displacement or offset has occurred along the dip direction. Note that in looking at the displacement on any fault we don't know which side actually moved or if both sides moved, all we can determine is the relative sense of motion.

    o Normal Faults - are faults that result from horizontal tensional stresses in brittle rocks and where the hanging-wall block has moved <u>down</u> relative to the footwall block.

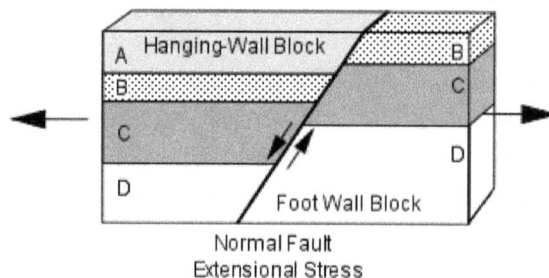

Normal Fault
Extensional Stress

*Horsts & Grabens* - Due to the tensional stress responsible for normal faults, they often occur in a series, with adjacent faults dipping in opposite directions. In such a case the down-dropped blocks form *grabens* and the uplifted blocks form *horsts*. In areas where tensional stress has recently affected the crust, the grabens may form *rift valleys* and the uplifted horst blocks may form linear mountain ranges. The East African Rift Valley is an example of an area where continental extension has created such a rift. The basin and range province of the western U.S. (Nevada, Utah, and Idaho) is also an area that has recently undergone crustal extension. In the basin and range, the basins are elongated grabens that now form valleys, and the ranges are uplifted horst blocks.

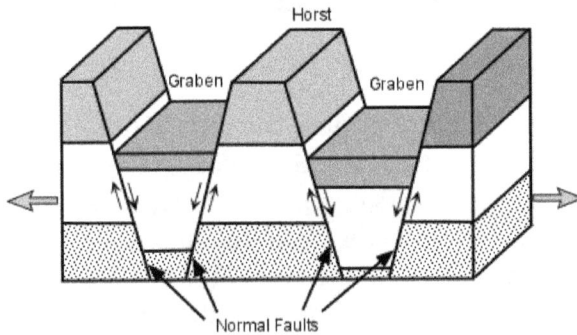

Half-Grabens - A normal fault that has a curved fault plane with the dip decreasing with depth can cause the down-dropped block to rotate. In such a case a half-graben is produced, called such because it is bounded by only one fault instead of the two that form a normal graben.

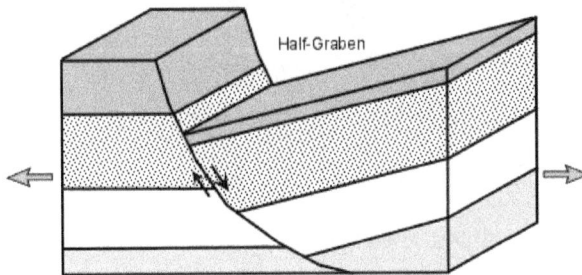

o   *Reverse Faults* - are faults that result from horizontal compressional stresses in brittle rocks, where the hanging-wall block has moved <u>up</u> relative the footwall block.

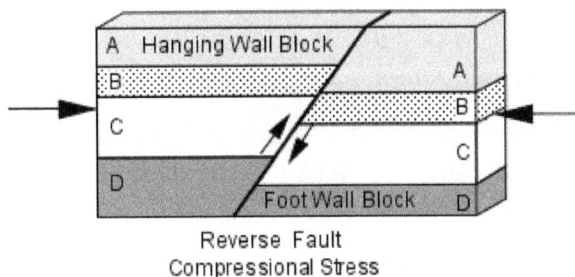

Reverse Fault
Compressional Stress

A *Thrust Fault* is a special case of a reverse fault where the dip of the fault is less than 45°. Thrust faults can have considerable displacement, measuring hundreds of kilometers, and can result in older strata overlying younger strata.

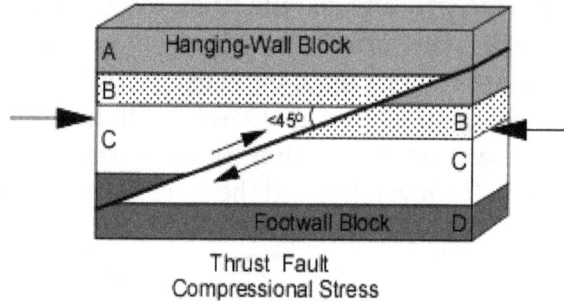

Thrust Fault
Compressional Stress

- *Strike Slip Faults* - are faults where the relative motion on the fault has taken place along a horizontal direction. Such faults result from shear stresses acting in the crust. Strike slip faults can be of two varieties, depending on the sense of displacement. To an observer standing on one side of the fault and looking across the fault, if the block on the other side has moved to the left, we say that the fault is a *left-lateral strike-slip fault*. If the block on the other side has moved to the right, we say that the fault is a *right-lateral strike-slip fault*. The famous San Andreas Fault in California is an example of a right-lateral strike-slip fault. Displacements on the San Andreas fault are estimated at over 600 km.

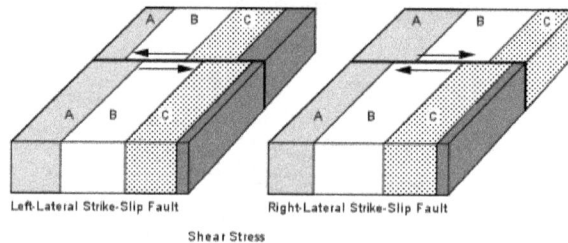

Left-Lateral Strike-Slip Fault          Right-Lateral Strike-Slip Fault

Shear Stress

## Evidence of Movement on Faults

Since movement on a fault involves rocks sliding past each other there may be left evidence of movement in the area of the fault plane.

- *Fault Breccias* are crumbled up rocks consisting of angular fragments that were formed as a result of grinding and crushing movement along a fault. When the rock is broken into clay or silt size particles as a result of slippage on the fault, it is referred to as *fault gouge*.

- *Slickensides* are scratch marks that are left on the fault plane as one block moves relative to the other. Slickensides can be used to determine the direction and sense of motion on a fault.

- *Mylonite* - Along some faults rocks are sheared or drawn out by ductile deformation along the fault. This results in a type of localized metamorphism called dynamic metamorphism (also called cataclastic metamorphism. The resulting rock is a fine grained metamorphic rock show evidence of shear, called a mylonite. Faults that show such ductile shear are referred to as *shear zones*.

## Deformation of Ductile Rocks

When rocks deform in a ductile manner, instead of fracturing to form faults or joints, they may bend or fold, and the resulting structures are called *folds*. Folds result from compressional stresses or shear stresses acting over considerable time. Because the strain rate is low and/or the temperature is high, rocks that we normally consider brittle can behave in a ductile manner resulting in such folds.

Geometry of Folds - Folds are described by their form and orientation. The sides of a fold are called *limbs*. The limbs intersect at the tightest part of the fold, called the *hinge*. A line connecting all points on the hinge is called the *fold axis*. An imaginary plane that includes the fold axis and divides the fold as symmetrically as possible is called the *axial plane* of the fold.

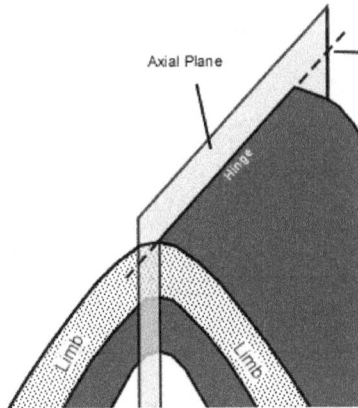

We recognize several different kinds of folds.

*Monoclines* are the simplest types of folds. Monoclines occur when horizontal strata are bent upward so that the two limbs of the fold are still horizontal.

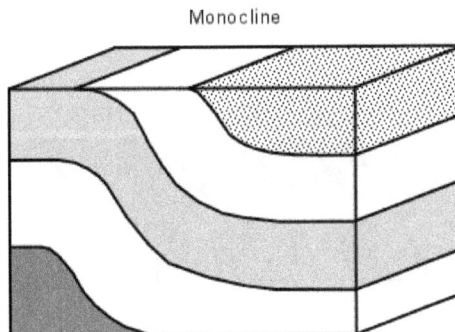

*Anticlines* are folds where the originally horizontal strata has been folded upward, and the two limbs of the fold dip away from the hinge of the fold.

Anticline

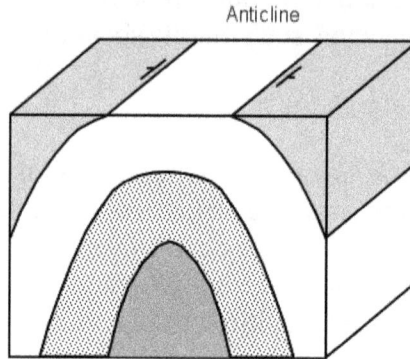

*Synclines* are folds where the originally horizontal strata have been folded downward, and the two limbs of the fold dip inward toward the hinge of the fold. Synclines and anticlines usually occur together such that the limb of a syncline is also the limb of an anticline.

Syncline

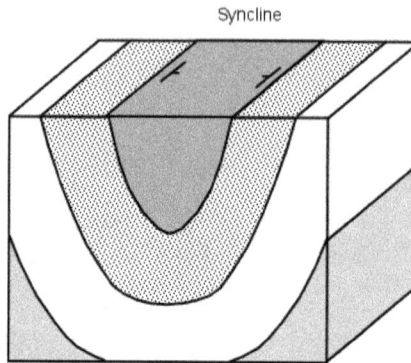

- In the diagrams above, the fold axes are horizontal, but if the fold axis is not horizontal the fold is called a *plunging fold* and the angle that the fold axis makes with a horizontal line is called the *plunge* of the fold.

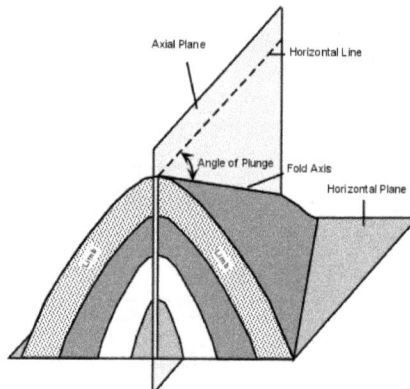

Note that if a plunging fold intersects a horizontal surface, we will see the pattern of the fold on the surface.

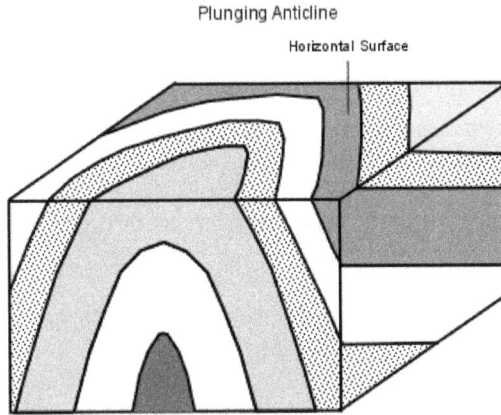

Plunging Anticline

Horizontal Surface

*Domes* and *Basins* are formed as a result of vertical crustal motion. Domes look like an overturned bowl and result from crustal upwarping. Basins look like a bowl and result from subsidence.

Folds are described by the severity of folding. an *open fold* has a large angle between limbs, a *tight fold* has a small angle between limbs.

Further classification of folds include:

- If the two limbs of the fold dip away from the axis with the same angle, the fold is said to be a *symmetrical fold*.

- If the limbs dip at different angles, the folds are said to be *asymmetrical folds*.

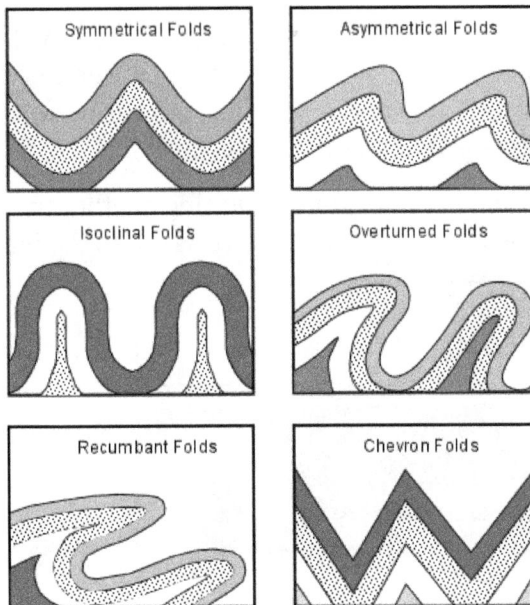

Symmetrical Folds

Asymmetrical Folds

Isoclinal Folds

Overturned Folds

Recumbant Folds

Chevron Folds

- If the compressional stresses that cause the folding are intense, the fold can close up and have limbs that are parallel to each other. Such a fold is called an *isoclinal fold* (iso means same, and cline means angle, so isoclinal means the limbs have the same angle). Note the isoclinal fold depicted in the diagram below is also a symmetrical fold.

- If the folding is so intense that the strata on one limb of the fold becomes nearly upside down, the fold is called an *overturned fold*.

- An overturned fold with an axial plane that is nearly horizontal is called a *recumbant fold*.

- A fold that has no curvature in its hinge and straight-sided limbs that form a zigzag pattern is called a *chevron fold*.

## Folds and Topography

Since different rocks have different resistance to erosion and weathering, erosion of folded areas can lead to a topography that reflects the folding. Resistant strata would form ridges that have the same form as the folds, while less resistant strata will form valleys.

## How Folds Form

Folds develop in two ways:

- *Flexural folds* form when layers slip as stratified rocks are bent. This results in the layers maintaining their thickness as they bend and slide over one another. These are generally formed due to compressional stresses acting from either side.

- *Flow folds* form when rocks are very ductile and flow like a fluid. Different parts of the fold are drawn out by this flow to different extents resulting in layers becoming thinner in some places and thicker in outer places. The flow results in shear stresses that smear out the layers.

- Folds can also form in relationship to faulting of other parts of the rock body. In this case the more ductile rocks bend to conform to the movement on the fault.

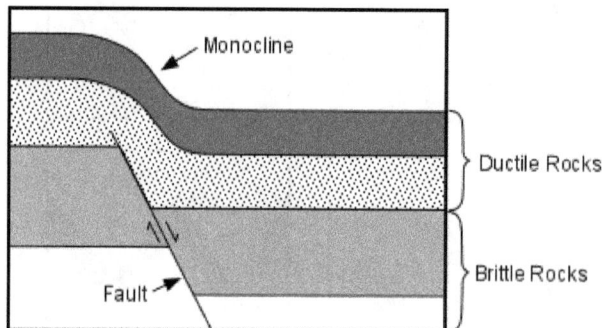

- Also since even ductile rocks can eventually fracture under high stress, rocks may fold up to a certain point then fracture to form a fault.

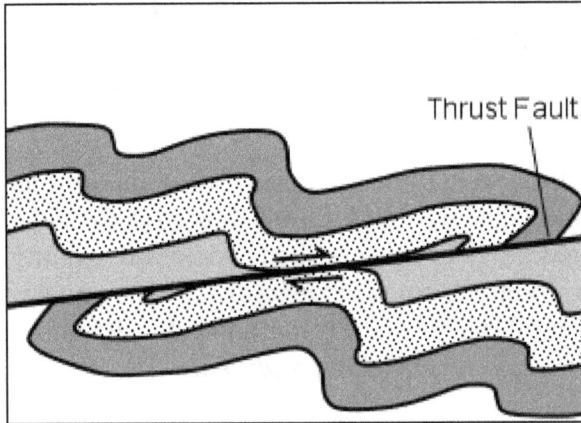

## Folds and Metamorphic Foliation

As we saw in our discussion of metamorphic rocks, foliation is a planar fabric that develops in rocks subject to compressional stress during metamorphism. It may be present as flattened or elongated grains, with the flattening occurring perpendicular to the direction of compressional stress. It also results from the reorientation, recrystallization, or growth of sheet silicate minerals so that their sheets become oriented perpendicular to the compressional stress direction. Thus, we commonly see a foliation that is parallel to the axial plane of the fold.

Shearing of rock during metamorphism can also draw out grains in the direction of shear.

## Mountains and Mountain Building Processes

One of the most spectacular results of deformation acting within the crust of the Earth is the formation of mountain ranges. Mountains frequently occur in elongate, linear belts. They are constructed by tectonic plate interactions in a process called orogenesis.

Mountain building (orogenesis) involves

- Structural deformation.
- Faulting.
- Folding.
- Igneous Processes.
- Metamorphism.
- Glaciation.

- Erosion.

- Sedimentation

Constructive processes, like deformation, folding, faulting, igneous processes and sedimentation build mountains up; destructive processes like erosion and glaciation, tear them back down again.

Mountains are born and have a finite life span. Young mountains are high, steep, and growing upward. Middle-aged mountains are cut by erosion. Old mountains are deeply eroded and often buried. Ancient orogenic belts are found in continental interiors, now far away from plate boundaries, but provide information on ancient tectonic processes. Since orogenic continental crust generally has a low density and thus is too buoyant to subduct, if it escapes erosion it is usually preserved.

## Uplift and Isostasy

The fact that marine limestones occur at the top of Mt. Everest, indicates that deformation can cause considerable vertical movement of the crust. Such vertical movement of the crust is called *uplift*. Uplift is caused by deformation which also involves thickening of the low density crust and, because the crust "floats" on the higher density mantle, involves another process that controls the height of mountains.

The discovery of this process and its consequences involved measurements of gravity. Gravity is measured with a device known as a gravimeter. A gravimeter can measure differences in the pull of gravity to as little as 1 part in 100 million. Measurements of gravity can detect areas where there is a deficiency or excess of mass beneath the surface of the Earth. These deficiencies or excesses of mass are called *gravity anomalies*.

A positive gravity anomaly indicates that an excess of mass exits beneath the area. A negative gravity anomaly indicates that there is less mass beneath an area.

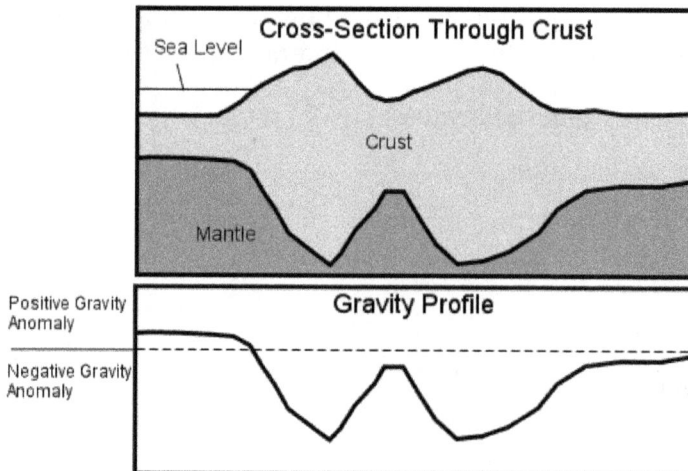

Negative anomalies exist beneath mountain ranges, and mirror the topography and crustal thickness as determined by seismic studies. Thus, the low density continents appear to be floating on higher density mantle.

The protrusions of the crust into the mantle are referred to as crustal roots. Normal crustal thickness, measured from the surface to the Moho is 35 to 40 km. But under mountain belts crustal thicknesses of 50 to 70 km are common. In general, the higher the mountains, the thicker the crust.

What causes this is the principal of *isostasy*. The principal can be demonstrated by floating various sizes of low density wood blocks in your bathtub or sink. The larger blocks will both float higher and extend to deeper levels in the water and mimic the how the continents float on the mantle.

It must be kept in mind, however that it's not just the crust that floats, it's the entire lithosphere. So, the lithospheric mantle beneath continents also extends to deeper levels and is thicker under mountain ranges than normal. Because the lithosphere is floating in the asthenosphere which is more ductile than the brittle lithosphere, the soft asthenosphere can flow to compensate for any change in thickness of the crust caused by erosion or deformation.

The Principle of isostasy states that there is a flotational balance between low density rocks and high density rocks. i.e. low density crustal rocks float on higher density mantle rocks. The height at which the low density rocks float is dependent on the thickness of the low density rocks. Continents stand high because they are composed of low density rocks (granitic composition). Ocean basins stand low, because they are composed of higher density basaltic and gabbroic rocks.

Isostasy is best illustrated by effects of glaciation. During an ice age crustal rocks that are covered with ice are depressed by the weight of the overlying ice. When the ice melts, the areas previously covered with ice undergo uplift.

Mountains only grow so long as there are forces causing the uplift. As mountains rise, they are eroded. Initially the erosion will cause the mountains to rise higher as a result of isostatic compensation. But, eventually, the weight of the mountain starts to depress the lower crust and sub-continental lithosphere to levels where they start to heat up and become more ductile. This hotter lithosphere will then begin to flow outward away from the excess weight and the above will start to collapse.

The hotter rocks could eventually partially melt, resulting in igneous intrusions as the magmas move to higher levels, or the entire hotter lower crust could begin to rise as a result of their lower density. These processes combined with erosion on the surface result in *exhumation*, which causes rocks from the deep crust to eventually become exposed at the surface.

## Causes of Mountain Building

There are three primary causes of mountain building.

1. Convergence at convergent plate boundaries.

   When oceanic lithosphere subducts beneath continental lithosphere magmas generated above the subduction zone rise, intrude, and erupt to form *volcanic mountains*. The compressional stresses generated between the trench and the volcanic arc create *fold-thrust mountain belts,* and similar compression behind the arc create a fold-thrust belt resulting in mountains. Mountains along the margins of western North and South America, like the Andes and the Cascade range formed in this fashion.

   Island arcs off the coast of continents can get pushed against the continent. Because of their low density, they don't subduct, but instead get accreted to the edge of the continent. Mountain ranges along the west coast of North America formed in this fashion.

2. Continental Collisions.

   Plate tectonics can cause continental crustal blocks to collide. When this occurs the rocks between the two continental blocks become folded and faulted under compressional stresses and are pushed upward to form *fold-thrust mountains*. The Himalayan Mountains (currently the highest on Earth) are mountains of this type and were formed as a result of the Indian Plate colliding with the Eurasian plate. Similarly the Appalachian Mountains of North America and the Alps of Europe were formed by such processes.

3. Rifting

   Continental Rifting occurs where continental crust is undergoing extensional deformation. This results in thinning of the lithosphere and upwelling of the asthenosphere which results in uplift. The brittle lithosphere responds by producing normal faults where blocks of continental lithosphere are uplifted to form grabens or half grabens. The uplifted blocks are referred to a *fault-block mountains.*

   The Basin and Range province in the western United states formed in this manner, including the Sierra Nevada on its western edge and the Grand Tetons in Wyoming.

## Cratons and Orogens

The continents can be divided into two kinds of structural units

- *Cratons* form the cores of the continents. These are portions of continental crust that have attained isostatic and tectonic stability and have cooled substantially since their formation. They were formed and were deformed more than a billion years ago and are the oldest parts of the continents. The represent the deep roots of former mountains and consist of metamorphic and plutonic igneous rocks, all showing extensive evidence of deformation.

- *Orogens* are broad elongated belts of deformed rocks that are draped around the cratons. They appear to be the eroded roots of former mountain belts that formed by continent - continent collisions. Only the youngest of these orogens still form mountain ranges.

The observation that the orogens are generally younger towards the outside of any continent suggests that the continents were built by collisions of plates that added younger material to the outside edges of the continents, and is further evidence that plate tectonics has operated for at least the last 2 billion years.

## Strike and Dip

A standard Brunton compass, used commonly by geologists for strike and dip measurements

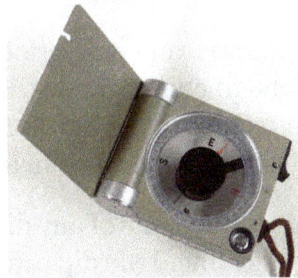

Stratum compass to measure dip and dip direction in one step

Strike and dip refer to the orientation or *attitude* of a geologic feature. The *strike line* of a bed, fault, or other planar feature, is a line representing the intersection of that feature with a horizontal plane. On a geologic map, this is represented with a short straight line segment oriented parallel to the strike line. *Strike* (or strike angle) can be given as either a quadrant compass bearing of the strike line (N25°E for example) or in terms of east or west of true north or south, a single three digit number representing the azimuth, where the lower number is usually given (where the example of N25°E would simply be 025), or the azimuth number followed by the degree sign (example of N25°E would be 025°).

The *dip* gives the steepest angle of descent of a tilted bed or feature relative to a horizontal plane, and is given by the number (0°-90°) as well as a letter (N,S,E,W) with rough direction in which the bed is dipping downwards. One technique is to always take the strike so the dip is 90° to the right of the strike, in which case the redundant letter following the dip angle is omitted (right hand rule, or RHR). The map symbol is a short line attached and at right angles to the strike symbol pointing in the direction which the planar surface is dipping down. The angle of dip is generally included on a geologic map without the degree sign. Beds that are dipping vertically are shown with the dip symbol on both sides of the strike, and beds that are level are shown like the vertical beds, but with a circle around them. Both vertical and level beds do not have a number written with them.

Another way of representing strike and dip is by dip and dip direction. The dip direction is the azimuth of the direction the dip as projected to the horizontal (like the trend of a linear feature in trend and plunge measurements), which is 90° off the strike angle. For example, a bed dipping 30° to the South, would have an East-West strike (and would be written 090°/30° S using strike and dip), but would be written as 30/180 using the dip and dip direction method.

Strike and dip are determined in the field with a compass and clinometer or a combination of the two, such as a Brunton compass named after D.W. Brunton, a Colorado miner. Compass-clinometers which measure dip and dip direction in a single operation (as pictured) are often called "stratum" or "Klar" compasses after a German professor. Smartphone apps are also now available, that make use of the internal accelerometer to provide orientation measurements. Combined with the GPS functionality of such devices, this allows readings to be recorded and later downloaded onto a map.

Any planar feature can be described by strike and dip. This includes sedimentary bedding, faults and fractures, cuestas, igneous dikes and sills, metamorphic foliation and any other planar feature in the Earth. Linear features are measured with very similar methods, where "plunge" is the dip angle and "trend" is analogous to the dip direction value.

Apparent dip is the name of any dip measured in a vertical plane that is not perpendicular to the strike line. True dip can be calculated from apparent dip using trigonometry if you know the strike. Geologic cross sections use apparent dip when they are drawn at some angle not perpendicular to strike.

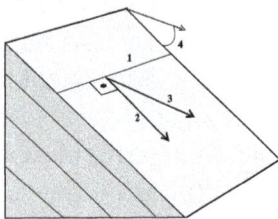

Strike and dip of the beds.
1-Strike, 2-Dip direction,
3-Apparent dip 4-Angle of dip

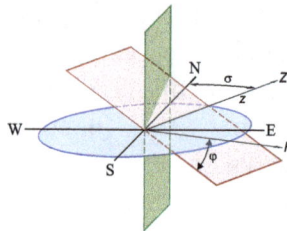

Strike and dip

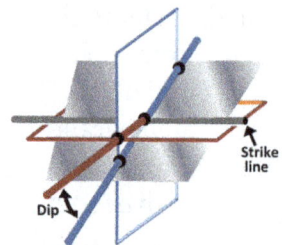

Strike line and dip of a plane describing attitude relative to a horizontal plane and a vertical plane perpendicular to the strike line

# Dip and Strike

Initially the sedimentary rocks are deposited on flat or gently inclined surfaces. Post depositional deformation is caused by the movements due to tectonic forces acting on rock body or mass which causes the tilting of the beds.

Dip: It is defined as inclination of the bedding plane with respect to horizontal. It is measured in a vertical plane lying at right angles to the strike of the bedding.

Strike: It is the geographical direction of a line produced by intersection between inclined layers and a horizontal plane (parallel to surface to Earth).

Direction of dip: It is the geographical direction, along which a bed has maximum slope.

Amount of dip: It is the angle which varies from oo to 90°, according to the inclination of the bed.

As discussed above the disposition of the beds (Figure) can be:

- Horizontal: angle of dip=0° (no dip)

- Inclined or tilted: angle of dip varying between 0°-90°

- Vertical: angle of dip=90°

Dip is the vector quantity as it has both direction and amount. Amount of dip denotes the angle of the bed inclination with respect to the horizontal. Strike is a scalar quantity because it has only direction and no magnitude. Strike of the bed is independent of its amount of dip.

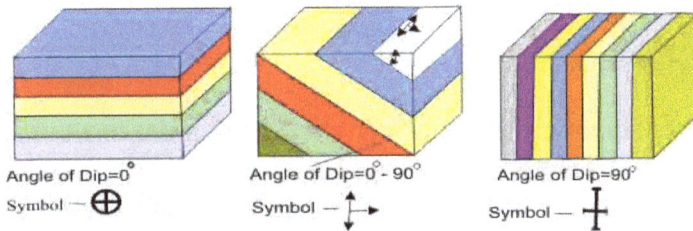

Horizontal, inclined, and vertical beds with symbolic representation on map

True and apparent dip: Dip can be of two types (Figure). Let us read about them.

- True dip: The maximum amount of inclination or slope of bed along a line perpendicular to the strike is the maximum slope with respect to the horizon, it is called true dip.

- Apparent dip: The dip of the bed measured in any direction other than that of true dip is called apparent dip. The amount of apparent dip is always less than amount of true dip.

Relation between dip and strike: The direction of dip and strike of any inclined or tilted bed must lie at right angles to each other. Thus true dip is in the direction perpendicular to strike. While mentioning the attitude of any inclined bed, dip amount and dip and strike direction should be mentioned.

Importance of Strike and Dip: Let us discuss the importance of strike and dip in structural geology.

(a) Determination of the younger bed or formation: In geological formations the older rocks deposit at the base which is superimposed by the younger rocks. Hence, in a tilted rock sequence when we move in the direction of dip then relatively beds of younger age will be encountered and vice-versa.

(b) Classification of geological structures: Dip and strike data provides useful information in the classification of rock or geologic structures.

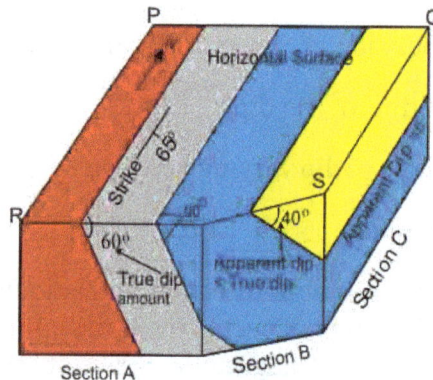

True dip, apparent dip and relation between dip and strike direction. PQRS is a horizontal plane. Beds are all dipping 60o towards east and they strike north-south

## Clinometer Compass

We have read that many rocks show planar structures in the form of bedding plane or foliation in any outcrop. They are not always horizontal, may show inclination of varying amounts. We have discussed that strike and dip describe the attitude of a rock layer in an outcrop. Let us recall that strike is the compass direction of a line formed by the intersection of a rock layer's surface with horizontal surface. Whereas the dip is measured at right angles to the strike and it is simply the amount of tilting or in other words the angle at which the rock layer is inclined from the horizontal. The amount of dip of a bedding plane varies. In order to understand the attitude and orientation of the rock structures, you are required to measure their dip and strike on the outcrops during the fieldwork. The measurement is done with the help of Clinometer Compass (Figure).

Clinometer Compass consists of a compass and clinometer. The compass has a circular dial which is anticlockwise graduated with 360 divisions in the outer most circle for reading the direction. This circle is called azimuthal circle; where 0 or 360 represents for North, 90 for East, 180 for South and 270 for West. It has a magnetic needle which rests in north- south direction of the Earth, when set free. The desired direction is read on the dial with help of N (north) marked end of this magnetic needle. One of the inner circles on the dial is also marked with directions such as E, NE, NNE, W etc. in supplement to the outermost circle to read the direction directly.

The inclination of a line with respect to horizontal is read with the help of clinometer fitted in compass (Figure). It has a frame with open space within the clinometer frame which allows seeing one of the inner circles in the dial marked with 0 to 900 graduation. The reading of inclination (or dip amount) which may be variable between horizontal (00) to vertical (900) is measured with the help of a mark in the middle of the frame opening. The bridge fitted outside the dial helps in taking reading of the inclination of bed. The bridge is movable which can rotate upto 900 angles from a plane parallel to the dial up to a plane perpendicular to the dial. The bridge line is always parallel to the North-South line of the dial. The locking device is present outside the dial ring which on pressing locks the magnetic needle as well as clinometer so that the reading may be noted even after removing the compass from outcrop.

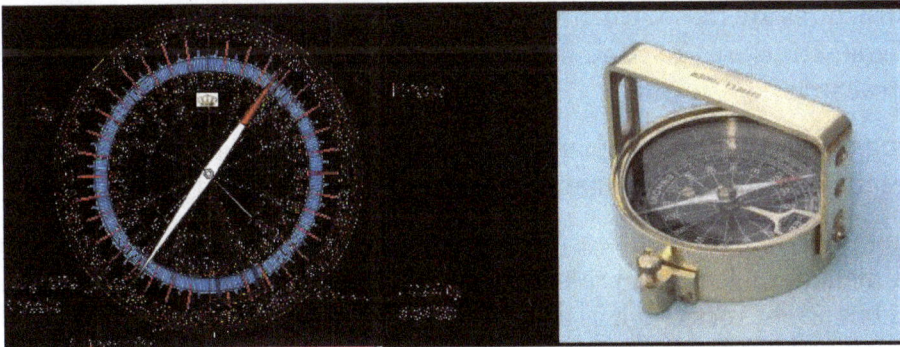

Sketch and photograph of clinometer compass

## Measuring Strike and Dip of an Inclined Bed

Let us discuss following procedure to measure the strike and dip of bed:

1. To measure the strike direction, place the compass on bedding plane in so that the bridge of the compass touches the bedding plane completely.

2. Then rotate the compass so as to ensure that the bridge becomes horizontal and one end of the bridge still touches the bedding plane. Now let the magnetic needle move freely. Let the needle come to rest and then read both ends of the azimuthal circle which represent strike.

3. In order to measure the amount of dip of the bed, draw a line on the bed perpendicular to strike and keep the bridge on the bedding plane along this line in such a way that the dial plane is vertical. The reading in clinometer gives the amount of dip.

4. In order to measure the direction of the dip, place the bridge along the line drawn on

the bedding plane so that the dial face the sky. Then rotate the compass to horizontal so that the bridge and N-S line of the dial both remain parallel to the line. The crown is often marked as N. Take care the crown in the dial is towards the dip direction of the bed.

5. On rotation of clinometer to the horizontal, the N marked end of the magnetic needle gives the dip direction. You can check your reading. The true dip and strike directions are always perpendicular to each other.

## Rock Structures

Rock structures can be broadly divided into:

- Planar structures;
- Linear structures;
- Folds and faults

## Planar Structures

Planar structures are those structures which are found as a plane or as a surface in the rocks. Planar structures go hand in hand with linear structures. Let us discuss major planar structures:

1. Bedding plane: They are the planes that bound a sedimentary bed. A bedding plane separates the two beds in a sedimentary rock.

2. Metamorphic foliation: It is a planar structure which is formed during deformation and metamorphism of a pre-existing rock. They are characteristic feature found in the metamorphic rocks such as slate, schist and gneiss.

3. Igneous foliation: They are observed with igneous rocks and are also known as primary foliation.

4. Fault plane: Fault plane is a fracture along which the rock blocks have been displaced.

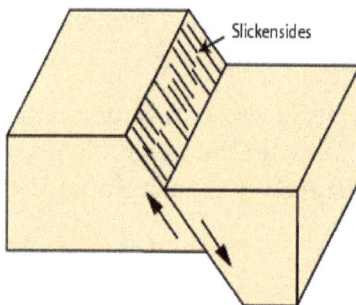

(a)                                                                                        (b)

(a) Displacement has taken place along the fault plane. Direction of movement is marked by arrows in opposite directions, and (b) Field photograph of slickensides

5. Bedding plane: They are the planes that bound a sedimentary bed. A bedding plane separates the two beds in a sedimentary rock.

6. Metamorphic foliation: It is a planar structure which is formed during deformation and metamorphism of a pre-existing rock . They are characteristic feature found in the metamorphic rocks such as slate, schist and gneiss.

7. Igneous foliation: They are observed with igneous rocks and are also known as primary foliation.

8. Fault plane: Fault plane is a fracture along which the rock blocks have been displaced.

9. Joints: They are irregular or regular planar separated portion found in the rocks. Joints are the fracture surfaces along which movement is negligible and/or not observable. They are among the most common of all geological features. The systematic study of joints in an area can unravel the timing and sequence of its formation. The cooling of lava or magma can produce joints in the rocks. Fracture and joints are among the most important geological structures considered in hydrology, engineering, mining projects.

Rhombohedral joints in igneous rock

## Linear Structures

Let us discuss about the linear structures. As the name suggests these structures is line like features observed in the rocks.

1. Mineral lineation: If the longer dimensions of the minerals are aligned in a particular direction, it gives rise to mineral lineation.

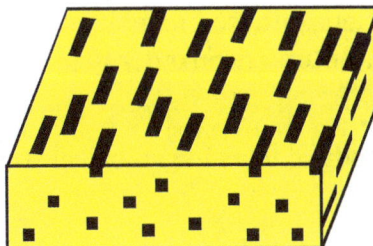

Mineral lineation

2. Crenulation lineation: They form when any planar surface such as foliation is affected by folding on small or microscopic scales so that hinge lines of these folds are aligned in a particular direction.

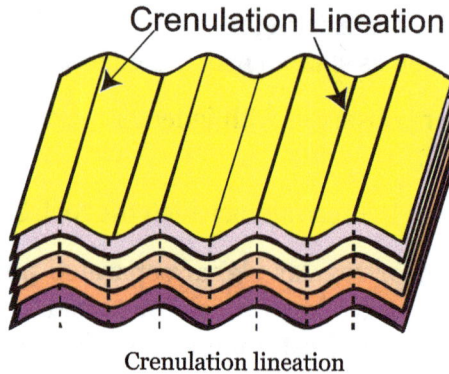

Crenulation lineation

3. Slickensides: They are lines like features developed on the fault plane because of friction generated between two faulted blocks and usually show polished surface.

4. Boudinage, pinch and swell and roddings: When the competent rock layers stretch and deforms into segments they form boudins. Individual boudins are commonly much longer in one dimension than other two. Roddings describes elongated mineral aggregates.

## Folds and Faults

Now we will discuss in detail about folds and faults

## Folds

Folds are combination of planar and linear structures. You can imagine folds in rocks are like folds in clothing. Similarly when the layers of rocks suffers gradual compression by the tectonic forces in the crust, they are pushed into folds. The layers of rocks can be crumpled or buckled into folds. Folding can be defined as the bending of rock strata due to compressional forces acting tangentially or horizontally towards a common point or plane from the opposite sides. We can define fold as the wave-like undulation wherein the bending or arching of the rock layers takes place due to forces of the Earth. Folding is a common form of deformation displayed in the layered rocks. Folds are best displayed by stratified formations, *viz.*, sedimentary or volcanic rocks or metamorphic rocks. Folds vary in size from centimetre on outcrop scale to hundreds of kilometers on the regional scale.

Importance of folding: The study of folds in geological studies is important because of following reasons:

• Folding exposes the deep seated rocks on the surface of the Earth.

- It increases the mineral deposits because of repetition of layers due to folding in a limited area.

- It facilitates development of site for deposition of mineral bearing solution.

- Folds serve as good host for oil and natural gas.

- Folding causes beautiful landscapes to develop which may enhance geotourism.

Parts of the fold: Now let us discuss the different parts or elements of a fold.

- Wavelength of fold can be defined as the minimum distance between its two successive points of same phase. Alternatively it can also be explained as the distance between two alternating inflection points (Figure).

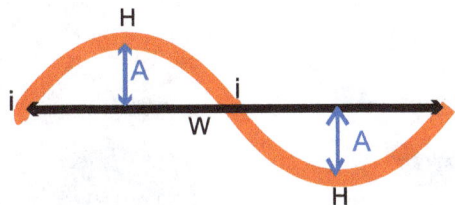

Parts of a fold (Abbreviations used: W-wavelength, A amplitude, H-hinge point, i-inflexion point

- Crest is the highest point in the profile section of the fold (Figure).

- Trough is the lowest point in the profile section of a fold (Figure).

- Crestal line is highest located line in the fold. It can be obtained by joining the crestal points in a folded layer (Figure).

- Trough line is obtained by joining the trough points of a folded layer (Figure). This line is located lowest in the fold.

- Culmination is the highest point located on the crestal line (Figure).

- Depression is the lowest point located on the trough line (Figure).

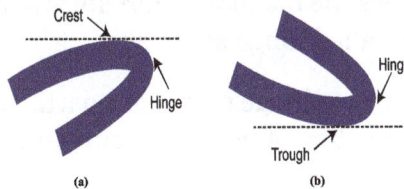

Figure 10(a) Crest, and (b) Trough of the fold

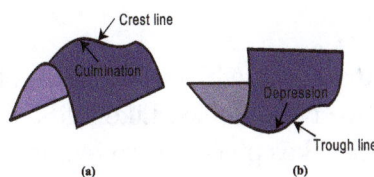

(a) Culmination, and (b) Depression

- Fold axis is an imaginary line which by moving parallel to itself generates the fold. The hinge line of a fold is considered equivalent to 'fold axis' if it is straight and the fold is cylindrical in nature.

- Inflexion point is that point where the fold limb changes its attitude.

- Inflexion line is obtained by joining the inflexion points of a folded layer.

- Limb is a side between inflexion point and hinge.

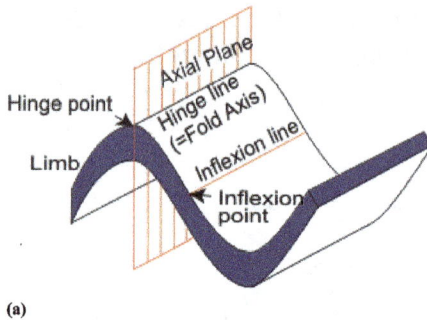

Parts of fold (a) Sketch, and (b) Field photograph of a fold

- Axial surface/ plane is formed by joining of fold hinge lines of successive beds. When the axial surface forms a plane it is called axial plane is formed by the axial surface.

- Amplitude of a fold is the length of perpendicular drawn from hinge point of the fold on the line joining the two inflection points of the fold.

- Hinge point It is the point of maximum curvature on the profile section of a fold. The profile of the fold is a cross section or transverse section across the hinge line of the fold.

- Hinge Zone: Sometimes the maximum curvature of the fold is not at a point but a set/group in a zone called hinge zone.

- Hinge line: It is the locus of hinge points of a particular bedding plane. Hinge or hinge line of a fold is the line of maximum curvature in the folded bed.

## Faults

Now let us discuss about faults.

Fault can be defined as a fracture along which there is an observable amount of dislocation or displacement of the two rock blocks. Like folds, faults also occur in all sizes. The development of fault in rock takes place due to tectonic stresses such as tensional, tangential or compressional or in combination.

Importance of Faults: Faults play a significant role in geological studies and can pose challenge to geologists while mapping.

- Faulting exposes the rocks from the deeper level to the Earth's surface which provides knowledge of the subsurface geology.

- Faults provide the excellent channel for the movement of mineralized solution or petroleum. They trap petroleum from migration and loss.

- Faults may create beautiful landscapes which enhance geotourism.

- They may be the cause of origin of earthquakes.

- Fault locations are very important prospects in mining and exploration.

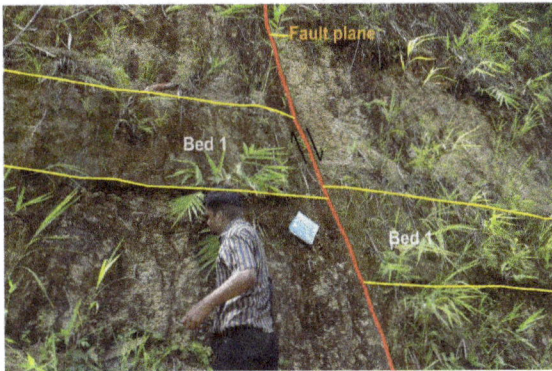

Field photograph of fault

Parts of Fault: Let us analyse different parts of fault.

- Fault surface/ fault plane: It is a surface or plane along which the dislocation takes place (Figure).

- Fault zone: It is a zone of numerous small scale fractures constituting fault (Figure).

- Fault line/ fault trace: It is the line formed by intersection of a fault with surface of the Earth or any given surface (Figure).

- Dip of the fault plane: It is the angle of inclination of the fault plane with respect to horizontal. The true dip is measured on a vertical section perpendicular to strike (Figure).

- Strike of the fault plane: It is the direction in which the fault plane cuts the horizontal ground and is strike of that plane (Figure).

- Hade of the fault plane: It is the angle made by fault plane with respect to a vertical line (Figure).

(a)
(b)
Figure 14(a)
Fault plane,
and (b) Fault
zone

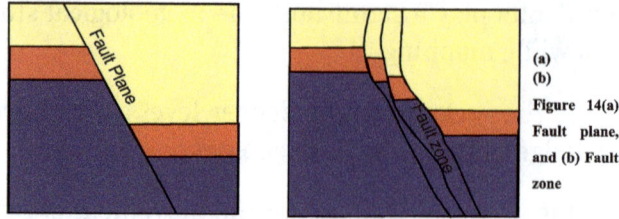

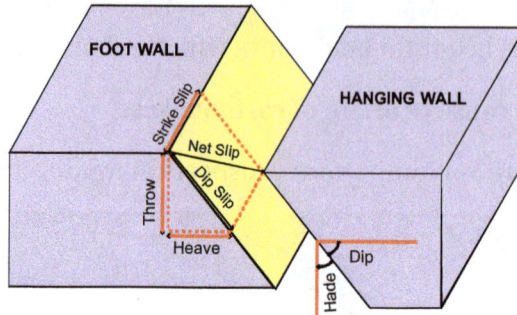

Parts of the fault3

- Hanging wall: The rock mass resting above the inclined fault plane is known as hanging wall (Figure).

- Foot wall: The rock mass resting below the inclined fault plane is known as foot wall (Figure).

- Slip: It refers to any displacement parallel to fault plane in the given direction (Figure).

- Net- Slip: It is the total amount of displacement measured parallel to fault plane (Figure).

- Dip-slip: It is the movement parallel to dip direction of the fault plane (Figure).

- Strike-slip: The movement parallel to strike of the fault plane (Figure).

- Oblique-slip: It is the movement called oblique-slip, if both dip-slip and strike-slip movements are present. (Figure).

- Heave: It is the horizontal component of the dip slip (Figure).

- Throw: It is the vertical component of the dip slip (Figure).

Recognition of Faults in the Field: Let us discuss features that are characteristically associated with fault plane. The presence of these features in the field is indicative of faulting.

- Visible displacements of veins, dikes, strata, *etc.* in the field are best evidence of faulting generally seen on the outcrop scale. For larger scale faulting geologists look for indirect evidences in the field.

- Presence of slickensides or fault breccia (contains angular fragments) along the fault plane is indicative of fault in the field.

- Dragging of strata is observed; there is some flexure or bending of beds present along a plane which may be a fault plane.

- Presence of crushing, shearing and pulverization of rock indicates faulting.

- Presence of mylonites (fine crushed rock materials) is indicative of fault zone.

- Silicification and mineralization is often associated with fault plane.

- Sedimentary or metamorphic rock types which are usually found at distant places, if they are juxtaposed together, presence of a fault may be indicated.

- Faulting often causes abrupt change in the topography, which can be observed in the form of cliff, triangular facets on mountains.

- If a river or stream shows an abrupt change in its flow direction or it makes waterfall; it is indicative of presence a fault.

- If the springs are found along a line, there may be presence of fault.

# References

- B.A. van der Pluijm and S. Marshak (2004). Earth Structure - An Introduction to Structural Geology and Tectonics (2nd ed.). New York: W. W. Norton. p. 656. ISBN 0-393-92467-X

- Livaccari, Richard F.; Burke, Kevin; Scedilengör, A. M. C. (1981). "Was the Laramide orogeny related to subduction of an oceanic plateau?". Nature. 289 (5795): 276–278. Bibcode:1981Natur.289..276L. doi:10.1038/289276a0

- V. Guerriero; et al. (2009). "Quantifying uncertainties in multi-scale studies of fractured reservoir analogues: Implemented statistical analysis of scan line data from carbonate rocks". Journal of Structural Geology. Elsevier. 32 (9): 1271–1278. Bibcode:2010JSG....32.1271G. doi:10.1016/j.jsg.2009.04.016

- BLATT, H.; TRACY, R.J. & OWENS, B.E.; 2006: Petrology, Igneous, Sedimentary, and Metamorphic, W.H. Freeman & company, New York (3rd ed.), ISBN 978-0-7167-3743-8

- "Digital Cartographic Standard for Geologic Map Symbolization". FGDC Geological Data Subcommittee. USGS. August 2006. Retrieved 20 March 2010

- V. Guerriero; et al. (2011). "Improved statistical multi-scale analysis of fractures in carbonate reservoir analogues". Tectonophysics. Elsevier. 504: 14–24. Bibcode:2011Tectp.504...14G. doi:10.1016/j.tecto.2011.01.003

# 4

# Minerals: Properties and Formation

Any chemical compound, which is naturally occurring and crystalline in nature, is called a mineral. Unlike rocks, minerals have specific chemical compositions and their study is referred to as mineralogy. This chapter discusses the various properties of minerals like their color, streak, hardness, crystal form, etc. as well as the geological processes of their formation.

## Mineralogy

Mineralogy is the branch of science which deals with the physico-chemical study of naturally occurring solid and crystalline materials. We can also say that it is the scientific study of all aspects of minerals concerning with conditions of formation during their origin and natural distribution. Internal structure of crystal, physical properties and chemical composition of minerals all are included in mineralogy.

A mineral is a homogeneous naturally occurring solid substance formed by inorganic processes having a definite but not fixed chemical composition with an ordered atomic arrangement.

### Chemical Composition of Minerals

Elements are the building blocks of minerals. Geochemical factors such as element abundance, solid solution limits, mineral stability, place a limit on the composition and stability of naturally occurring compounds, hence, there are relatively less number of minerals. Minerals and synthetic compounds may have identical structures. However, they differ in the fact that minerals are rarely pure substances and typically show wide variation in their composition. According to the composition, variation of minerals ranges from pure elements (Fe, Au, Ag) with relatively simple compounds (eg. PbS galena, KCI- sylvite) to very complex compounds (eg. Steenstrupine). Chemically simple minerals ($SiO_2$) do not necessarily have simple structures, e.g., alpha quartz. Chemical classifications of minerals are based on the predominant anion or anionic group. The following classes are recognized: (1) native elements; (2) sulphide, telluride, arsenide and selenide minerals including sulphosalts of antimony and bismuth; (3) halides; (4) oxides; (5) hydroxides; (6) carbonates; (7) nitrates; (8) borates; (9) chromates; (10) tungstates; (11) molybdates; (12) phosphates; (13) arsenates; (14) vanadates; (15) silicates and aluminosilicates. Because of the dominance of oxygen, silicon and aluminum in the earth, silicates and aluminosilicates are quantitatively the most important class of minerals. Minerals of mixed anion composition, e.g. F in topaz or apatite, S in pyrite, OH in talc and mica, Cl in biotite, are usually classified according to the nature of the dominant anion. Hybrid minerals, such as valleriite and tochilinite, are sulphides containing layers of hydroxides, and are not common. Each major compositional class of minerals is subdivided into groups of minerals having similar crystal structures (Table).

Table : Classification of Minerals

| **Mineral Group** | **Anionic  or Anionic Complex** | **Representative Minerals** |
|---|---|---|
| Native element | - | Sulfur, gold, silver, copper, diamond graphite |
| Sulfides | $S2-$ | Pyrite, galena, sphalerite, chalcopyrite |
| Oxides | $O2-$ | Hematite, magnetite, chromite |
| Halides | $Cl2-, F-$ | Halite, fluorite |
| Sulfates | $(SO4) 2-$ | Anhydrite, gypsum, barite |
| Carbonates | $(CO3) 2-$ | Calcite, dolomite |
| Phosphates | $(PO4)3-$ | Apatite |
| Silicates | $(SiO4) 2-$ | Quart, feldspar |

## Physical Properties

Calcite is a carbonate mineral ($CaCO_3$) with a rhombohedral crystal structure.

An initial step in identifying a mineral is to examine its physical properties, many of which can be measured on a hand sample. These can be classified into density (often given as specific gravity); measures of mechanical cohesion (hardness, tenacity, cleavage, fracture, parting); macroscopic visual properties (luster, color, streak, luminescence, diaphaneity); magnetic and electric properties; radioactivity and solubility in hydrogen chloride (HCl).

*Hardness* is determined by comparison with other minerals. In the Mohs scale, a standard set of minerals are numbered in order of increasing hardness from 1 (talc) to 10 (diamond). A harder mineral will scratch a softer, so an unknown mineral can be placed in this scale by which minerals it scratches and which scratch it. A few minerals such as calcite and kyanite have a hardness that depends significantly on direction. Hardness can also be measured on an absolute scale using a sclerometer; compared to the absolute scale, the Mohs scale is nonlinear.

Aragonite is an orthorhombic polymorph of calcite.

*Tenacity* refers to the way a mineral behaves when it is broken, crushed, bent or torn. A mineral can be brittle, malleable, sectile, ductile, flexible or elastic. An important influence on tenacity is the type of chemical bond (*e.g.*, ionic or metallic). Of the other measures of mechanical cohesion, *cleavage* is the tendency to break along certain crystallographic planes. It is described by the quality (*e.g.*, perfect or fair) and the orientation of the plane in crystallographic nomenclature. *Parting* is the tendency to break along planes of weakness due to pressure, twinning or exsolution. Where these two kinds of break do not occur, *fracture* is a less orderly form that may be *conchoidal* (having smooth curves resembling the interior of a shell), *fibrous*, *splintery*, *hackly* (jagged with sharp edges), or *uneven*.

If the mineral is well crystallized, it will also have a distinctive crystal habit (for example, hexagonal, columnar, botryoidal) that reflects the crystal structure or internal arrangement of atoms. It is also affected by crystal defects and twinning. Many crystals are polymorphic, having more than one possible crystal structure depending on factors such as pressure and temperature.

## Physical Properties of Mineral

The mineral can often be identified in the field using basic following properties:

1. Color
2. Streak
3. Hardness
4. Cleavage or fracture
5. Crystalline structure
6. Diaphaneity or amount of Transparency
7. Tenacity
8. Magnetism
9. Lustre
10. Odour
11. Taste and
12. Specific Gravity

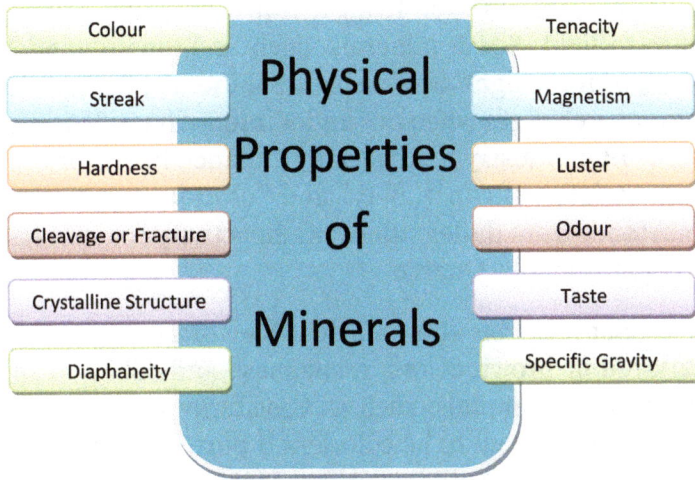

Properties of minerals

# Color

Idiochromatic minerals are said to be "self coloured" minerals because of their composition. The color of these minerals is caused by the presence of element(s) in the chemical composition of the mineral. e. g. Cu in azurite (blue), Cu in malachite (green), Mn in rhodonite and rhodochrosite (pink).

Allochromatic minerals are said to be "other colored" as impurities in traces are present in their composition or due to defects in their structure. In these minerals, the color is a variable and unpredictable property. e. g. blue in Amazonite (orthoclase), yellow in Heliodor (spodumene) and the rose in rose quartz.

Pseudochromatic minerals are "false colored" due to tricks in light diffraction. In these cases, color is variable but a unique property of the mineral. Examples are the colors produced by precious opal and the shiller reflections of labradorite.

Color is the most eye-catching feature of many minerals. Some minerals will always have a similar color, such as Gold, whereas some minerals, such as Quartz and Calcite, come in all colors. The presence and intensity of certain elements will determines a specific mineral's color. Minerals with an inherent color (i.e. all specimens of the mineral are the same color) have essential elements in them which cause their color. Good examples are Azurite and Malachite, which have their strong blue and green color due to their copper in their atomic structure. But there are many minerals which have slight additions of color-causing elements in some specimens that cause it to be a different color. For example, pure Quartz ($SiO_2$), is colorless, whereas Amethyst, a purple variety of quartz, has its purple color caused by traces of the element iron. The amount of iron present determines the intensity of the color.

Certain minerals exhibit a color change when exposed to light, heat, radiation, or when atomic anominalies are present. Red Realgar transforms into yellow Paraealgar upon repeated exposure to light. Some minerals, such as Proustite and Vivianite, darken upon prolonged exposure to light, whereas other minerals, such as Kunzite fade. Some minerals undergo color changes when put under intense heat. This method is commonly used in the gemstone industry to artificially enhance the color of many gemstones. For example, some varieties of Topaz, Beryl, and Corundum are heat treated to produce deep colored gemstones from duller stones. Radioactivity can also have an effect, as is the cause of the color of Smoky Quartz.

Most secondary copper minerals show a bright blue or green color. Iron is usually responsible for dark red or brown colors, manganese and cobalt for pink, and chromium for deep green. Some minerals, such as Cassiterite and Zincite, have a chemical structure that would cause them to be colorless if pure, but due to impurities that are always present, they are never found colorless. Most secondary uranium minerals exhibit bright neon yellow or green colors. Often, a mineral has a coating or has been pseudomorphed, causing it to exhibit the color of the replaced or coated mineral. Two common examples are a Hematite coating Quartz, which causes the Quartz to be red, and a Limonite pseudomorph after Pyrite.

Minerals composed mostly of the elements aluminum, sodium, and potassium are usually colorless or very lightly colored. In some cases, the color of a mineral may depend on its atomic bonding rather than composition, such as by Diamond and Graphite. Both these minerals have the same are formed from the same material (carbon), yet one is almost always white or very lightly colored, while the other is dark gray to black. Anominalies in the chemical structure of Halite are responsible for the deep blue and purple hues sometimes seen in this otherwise lightly colored mineral.

Inclusions of one mineral within a host mineral can also cause a color difference. Two fine examples are:

> 1) Chlorite inclusions in Quartz, causing the otherwise clear Quartz to be green.

> 2) Rutile inclusions in Quartz, which give the Quartz a golden hue.

Several minerals tarnish or oxidize, thereby affecting their color. The best examples are:

> 1) Silver, which tarnishes black, yellow, or brown.

> 2) Copper, which oxidizes green.

> 3) Bornite and Chalcopyrite, which oxidizes an iridescent array of colors.

Some minerals, such as Opal, display a multicolored effect when viewed from different angles. This is called opalescence, or "Play of Color". A few minerals appear to change color when viewed in different light. Alexandrite, a variety of Chrysoberyl, is usually dark green in natural light, but takes on a purplish hue in artificial light. Other minerals will change color when viewed at different angles. This is called dichroism or pleochroism. Cordierite, the most famous dichroic mineral is bluish-purple, but turns gray when rotated or viewed at a different angle. The properties of *opalescence*, *labradorescence*, and *dichroism*, as well as other optical properties are explained in greater detail in the section "Other Properties".

Mineral colors may be artificially enhanced in various ways, especially when they are used as gemstones. This includes heating and irradiation (both previously mentioned), as well as dying. A few minerals, especially Agate, are sometimes dyed to enhance their color.

## Play of Colors

One of the salient features of color mechanisms is the play-of-colors. Interference of light reflected from the surface or from within a mineral may cause the color of the mineral to change as the angle of incident light changes. This gives the mineral a beautiful iridescent quality and colors to the minerals for example, bornite ($Cu_5FeS_4$), hematite ($Fe_2O_3$), sphalerite ($ZnS$), opal and some specimens of labradorite (plagioclase). It is sometimes seen on tarnished sulfides like chalcopyrite or pyrite and occasionally on an oxide or a silicate mineral. It is caused by an optical effect called interference. When light strikes a shiny, transparent layer at an oblique angle the light rays are split - part of the light is reflected off the upper surface and part goes into the layer. Some of the light that enters the layer is reflected off the bottom surface and exists parallel to the other reflection. However, white light is composed of a numbers of colors, each of which travels at a different speed in the layer. So when the split light rays emerge from the layer they have been retarded and their wave patterns are out of synchronization with the part of the light reflected off the top of the layer. One or two colors will be in phase and the rest out of phase. The in phase color will be very bright and the out of phase colors will be very pale or absent. Thus, we may see a rich, glowing color coming from a colorless, transparent stone, but only in one special direction.

## Colours Due to Wavelength of Light

Minerals are colored because certain wavelengths of light are absorbed, and the mineral color then results from the combination of those wavelengths which reach the eye if light is not absorbed, the mineral is colorless in reflected or refracted light and is black if all waveof light are absorbed. Pleochroism may be present in certain minerals if light travels along crystallographic axes. A mineral may display more than one color when rotated or viewed at different angles. Cordierite specimens often display a clear-white and violet color when rotated indicating a dichroism quality.

## Colour Due to Impurities

Tiny amounts as little as one tenth of 1% of an impurity in the molecular structure of a mineral can determine that mineral's color. The amount and type of impurities affects the color of the mineral. Minerals with an inherent color have essential elements in them which cause their color. The examples are Azurite and Malachite, which have their strong blue and green color due to their copper in their atomic structure. But there are many minerals which have slight additions of color-causing elements in some specimens that cause it to be a different color. For example, pure Quartz ($SiO_2$), is colorless, whereas Amethyst, a purple variety of quartz, has its purple color caused by traces of the element iron. The amount of iron present determines the intensity of the color. Trace amounts of titanium or manganese turn quartz pink.

## Colour due to Charge Transfer

When two or more elements in a mineral exchange electrons then this is called charge transfer. The movement of electrons results in selective absorption of light. For Example,

Sapphire: Sapphires contain small amounts of titanium and iron. The electron transfer between Ti and Fe causes light in the yellow through red spectrum to be absorbed, producing the deep blue color sapphires.

Aquamarine: Small amounts of iron in valence states $Fe^{2+}$ and $Fe^{3+}$ cause an electron transfer that absorbs red light, resulting in the color blue.

Tourmaline: When manganese ($Mn^{2+}$) and titanium ($Ti^{4+}$) swap electrons, it creates a yellow-green color.

## Streak

Streak is the color of the mineral in powdered form. The streak of a mineral is quite different than the mineral. Although the colour of a mineral may vary, the streak is usually constant and is thus useful in mineral identification. Streak is a more accurate illustration of the mineral's color, streak is a more reliable property of minerals than color for identification.

The apparent color of a mineral can vary widely because of trace impurities or a disturbed macroscopic crystal structure. Small amounts of an impurity that strongly absorbs a particular wavelength can radically change the wavelengths of light that are reflected by the specimen, and thus change the apparent color. However, when the specimen is dragged to produce a streak, it is broken into randomly oriented microscopic crystals, and small impurities do not greatly affect the absorption of light.

The surface across which the mineral is dragged is called a "streak plate", and is generally made of unglazed porcelain tile. In the absence of a streak plate, the unglazed underside of a porcelain bowl or vase or the back of a glazed tile will work. Sometimes a streak is more easily or accurately described by comparing it with the "streak" made by another streak plate.

Because the trail left behind results from the mineral being crushed into powder, a streak can only be made of minerals softer than the streak plate, around 7 on the Mohs scale of mineral hardness. In case of harder minerals, the color of the powder can be determined by filing or crushing with a hammer a small sample, which is then usually rubbed on a streak plate. Most minerals that are harder have an unhelpful white streak.

Some minerals leave a streak similar to their natural color, such as cinnabar and lazurite. Other minerals leave surprising colors, such as fluorite, which always has a white streak, although it can appear in purple, blue, yellow, or green crystals. Hematite, which is black in appearance, leaves a red streak which accounts for its name, which comes from the Greek word "haima", meaning "blood." Galena, which can be similar in appearance to hematite, is easily distinguished by its gray streak.

Two minerals that have similar outward color may have different colors when powdered. For instance, the minerals hematite and galena can be confused when both have a gray color. However, hematite's streak is blood-red, while galena's streak is lead gray. Hematite is probably the most well known example of streak with its completely surprising streak color.

Unfortunately for collectors, translucent minerals have, usually, a rather undiagnostic white streak. Many opaque minerals similarly have a rather unhelpful black streak. However, there are about 20% of minerals that have unique shades of red, orange, yellow, blue, green, gray and even purple streaks and in many cases these streaks are very diagnostic.

There are many reasons why a mineral might have a different streak color than its outward color. First, there are translucent minerals that are colored by trace elements. These minerals require a large amount of travel time for light to pick up the coloring effects of these trace elements. As a result, small crystals are typically paler than large crystals and extrapolating down to a speck of powder will remove all coloring effects of a trace element and result in a white streak. A translucent mineral that has variable colors will almost certainly have a white streak.

Secondly, a mineral's structure and/or coatings can affect the outward color of a mineral and the streak in many ways is the true color of the mineral. Pyrite (known as "Fool's Gold") is always brassy yellow when found in crystals, even broken crystals, of any size; but when powdered, produces a black streak. It is the structure and chemistry of pyrite that produces the brassy yellow color, but only when enough structure is there. Gold's streak by the way is yellow! One note of caution, a streak plate is only about 6.5 in hardness and a mineral harder than 6.5 will not leave a streak on a streak plate but might scratch the plate leaving a white powder of porcelain, not the mineral! Fortunately most minerals harder than 6.5 have a white streak.

## Streak Test

The "streak test" is a method used to determine the color of a mineral in powdered form. The color of a mineral's powder is often a very important property for identifying the mineral.

The streak test is done by scraping a specimen of the mineral across a piece of unglazed porcelain known as a "streak plate." This can produce a small amount of powdered mineral on the surface of the plate. The powder color of that mineral known as its "streak."

| Streak Colors of Common Minerals | |
|---|---|
| **Andalusite** | White or colorless (hardness is about the same as the streak plate). |
| **Anhydrite** | White. |
| **Apatite** | White. |
| **Arsenopyrite** | Dark grayish black. |
| **Augite** | White to greenish gray. Augite can be splintery and close to the hardness of the streak plate, so brittle fragments, rather than a powder, will sometimes be produced. |
| **Azurite** | Light blue. |
| **Barite** | White. |
| **Bauxite** | White. Often discolored to pink, brown, or red by iron staining. |
| **Benitoite** | White. |
| **Beryl** | Colorless (harder than the streak plate). |
| **Biotite** | White to gray (don't be deceived by flakes). |
| **Bornite** | Grayish black. |
| **Calcite** | White. |
| **Cassiterite** | Colorless. |
| **Chalcocite** | Grayish black. |
| **Chalcopyrite** | Greenish black. |
| **Chlorite** | Greenish to greenish-black to white. |
| **Chromite** | Dark brown. |
| **Chrysoberyl** | Colorless (harder than the streak plate). |
| **Cinnabar** | Red. |
| **Clinozoisite** | White. |
| **Copper** | Metallic copper red. |
| **Cordierite** | Colorless (harder than the streak plate). |
| **Corundum** | Colorless (harder than the streak plate). |
| **Cuprite** | Brownish red. |
| **Diamond** | Colorless (harder than the streak plate). |
| **Diopside** | White to light green. |
| **Dolomite** | White. |
| **Enstatite** | White to gray. |

| | |
|---|---|
| **Epidote** | White or colorless (about the same hardness as the streak plate). |
| **Euclase** | White or colorless (when harder than the streak plate). |
| **Fluorite** | White. |
| **Fuchsite** | White (often sheds tiny green mica flakes). |
| **Galena** | Lead gray to black. |
| **Garnet** | Colorless (harder than the streak plate). |
| **Glauconite** | Dull green. |
| **Gold** | Metallic gold yellow. |
| **Graphite** | Black. |
| **Gypsum** | White. |
| **Halite** | White. |
| **Hematite** | Red to reddish brown. |
| **Hornblende** | White. Brittle, often leaves black cleavage debris behind instead of a streak. |
| **Ilmenite** | Black. |
| **Jadeite** | Colorless (harder than the streak plate). |
| **Kyanite** | White or colorless (about the same hardness as the streak plate in one direction). |
| **Limonite** | Yellowish brown. |
| **Magnesite** | White. |
| **Magnetite** | Black. |
| **Malachite** | Green. |
| **Marcasite** | Grayish Black. |
| **Molybdenite** | Bluish gray, grayish black. |
| **Monazite** | White. |
| **Muscovite** | White, often sheds tiny cleavage flakes. |
| **Nepheline** | White. |
| **Nephrite** | Colorless (harder than the streak plate). |
| **Olivine** | White or colorless (about the same hardness as the streak plate). Often sheds tiny granules instead of a powder. |
| **Orthoclase** | White. |
| **Plagioclase** | White. |
| **Prehnite** | White. |
| **Pyrite** | Greenish black to brownish black. |
| **Pyrophyllite** | White. |
| **Pyrrhotite** | Grayish black. |
| **Quartz** | Colorless (harder than the streak plate). |
| **Rhodochrosite** | White. |
| **Rhodonite** | White. |
| **Rutile** | Pale brown. |
| **Scapolite** | White. |
| **Serpentine** | White. |

| Siderite | White, very light brown. |
|---|---|
| Sillimanite | White or colorless (about the same hardness as the streak plate). |
| Silver | Silvery white. |
| Sodalite | White or light blue. |
| Sphalerite | White to yellowish brown, often with an odor of sulfur. |
| Spinel | Colorless (harder than the streak plate). |
| Spodumene | White or colorless (about the same hardness as the streak plate). |
| Staurolite | Colorless (harder than the streak plate). |
| Sulfur | Yellow. |
| Sylvite | White. |
| Talc | White to pale green. |
| Titanite | White. |
| Topaz | Colorless (harder than the streak plate). |
| Tourmaline | Colorless (harder than the streak plate). Specimens often fracture, shedding small particles. |
| Turquoise | White, greenish, bluish. |
| Uraninite | Brownish black, grayish. |
| Witherite | White. |
| Wollastonite | White. |
| Zircon | Colorless (harder than the streak plate). |
| Zoisite | White. |

## How to Conduct the Streak Test

The streak test should be done on clean, unweathered, or freshly broken specimens of the mineral. This is done to reduce the possibility that a contaminant, weathered coating, or tarnish will influence the results of the test.

The preferred method for conducting a streak test is to pick up a representative specimen of the mineral with the hand that you write with. Select a representative point or protrusion on the specimen that will be scraped across the streak plate. With your other hand, place the streak plate flat on a tabletop or laboratory bench. Then, while holding the streak plate flat and firmly in place on the tabletop, place the point of the specimen firmly against the streak plate, and, while maintaining firm pressure, drag the specimen across the plate. Now examine the streak to determine its color and to confirm that it is a powder, instead of grains, splinters, or broken pieces.

The most common error made by people who are doing the streak test for the first time is to lightly rub the specimen back and forth on the surface of the streak plate. This will not produce a proper streak. Some mineral specimens are so hard that very firm pressure and determination are required to produce a mineral powder.

## Why use the Streak Test?

The streak test is valuable because many minerals occur in a variety of apparent colors - but all specimens of that mineral share a similar streak color. For example: specimens of hematite can be black, red, brown, or silver in color and occur in a wide variety of habits; however, all specimens of hematite produce a streak with a reddish color. This is a valuable test for hematite. It can be used to differentiate hematite from a large number of other opaque minerals with a high specific gravity and similar color and habit.

Fluorite is another mineral where the apparent color can be different from the color of the streak. Specimens of fluorite can be green, yellow, purple, blue, or colorless. However, all specimens of fluorite have a white streak. Specimens of pyrite always have a brassy yellow color; however, all specimens of pyrite produce a black streak.

## Don't Be Deceived!

A number of things can cause a streak test to give unreliable results. To avoid problems, keep the following items in mind.

- Always do the streak test using a surface of the specimen that has not been weathered. Many weathered specimens are coated with a layer of alteration products that have a different streak color. If you are in doubt and are permitted to break the specimen, testing on a freshly broken surface is a good idea.

- Repeat the test using two different parts of the specimen or two different pieces of the same material for confirmation.

Contamination Alters Streak: This specimen of bauxite from Demerara, Guyana should have a white streak; however, it has a pinkish streak because it is contaminated by iron-staining. The streak also varies depending upon what part of the specimen is tested. Specimen is approximately 4 inches (10 centimeters) across.

- Be alert for contaminated specimens. For example: bauxite is sometimes contaminated with iron oxides that produce a streak that is not white in color.

- Some minerals are brittle or have a granular habit. When these are scraped across a streak plate, a trail of dislodged grains or broken pieces are produced instead of a powder. Rub the tip of your index finger across the streak plate to place a small amount of the mineral powder on your finger tip. Then rub the tip of your index finger against the tip of your thumb. A powder will have a smooth feel between your finger and thumb. Brittle fragments or granules will feel gritty. Streak color is determined from a powder rather than fragments.

- Streak plates usually have a Mohs hardness of between 6.5 and 7. Many minerals are harder than the streak plate. Instead of leaving a powder behind when dragged across a streak plate, they will scratch the streak plate or fracture into small pieces. Minerals that are harder than the streak plate are said to have "no streak" or a "colorless streak."

- If the results of your streak test seem inaccurate, be cautious. The streak test should be used as a "hint" leading to the identification of a mineral. Identification of a mineral should always be based upon observations of several different mineral properties.

## Refreshing your Streak Plate

Streak plates that have been used heavily will be covered with streaks and powdered mineral. They can easily be cleaned with water and a piece of wet or dry 220 grit sandpaper. Aluminum oxide or silicon carbide sandpaper works best because the granules are hard enough to smooth the surface of the streak plate. The sanding should be done wet to control dust.

The best way to learn about minerals is to study with a collection of small specimens that you can handle, examine, and observe their properties.

## Other Uses for Streak Plates

In addition to their use in doing the streak test, streak plates can be used any time you need a small amount of powdered mineral. In doing the acid test to distinguish calcite from dolomite, dolomite might require being powdered to show effervescence with dilute hydrochloric acid. Simply use the streak plate to make some powder of your specimen and add acid to it right on the streak plate. For this test, a black streak plate makes observation easier because powdered dolomite is white.

A few minerals will produce an odor upon being broken or powdered. For example, sphalerite releases an odor of sulfur when it is broken or powdered. Scraping it across a streak plate is a convenient way to conduct this test.

Hints to other mineral properties can be obtained while doing the streak test. Minerals harder than the streak plate are quickly identified. Experienced testers can estimate the hardness of a specimen by how difficult it is to mark the streak plate. Olivine often reveals its granular nature, augite often reveals its splintery cleavage, and black tourmaline often reveals its brittleness. When you do the streak test, look for more than the color of a specimen's powder.

## Hardness

Friedrich Mohs in 1812 developed a standard scale for calculating the extent of hardness. Hardness varies greatly in minerals. Its determination is one of the most important tests used in identification of minerals. Hardness is the resistance offered by smooth surface of a mineral on scratching. The hardness of a mineral might be said to be its "scratchability". Hardness, like many other physical properties depends upon atomic structure of mineral. It varies with density of packing in structure. On the basis of their relative hardness, minerals are ranked from 1 to 10. Softer minerals are scratched by harder minerals. Moh's scale is based on the ten index minerals while other minerals are ranked relative to these. For instance, a mineral having hardness of 6.5 can scratch feldspar but not quartz.

## Moh's Hardness Scale

| Talc | Soft |
| Gypsum | |
| Calcite | |
| Fluorite | |
| Apatite | |
| Feldspar | |
| Quartz | |
| Topaz | |
| Corundum | |
| Diamond | Hardest |

Hardness plays a major role in identifying a mineral. It can make the identification process much simpler by considerably narrowing a search.

Hardness is defined by how well a substance will resist scratching by another substance. For example, if mineral A scratches mineral B, and mineral B does not scratch mineral A, then mineral A is harder than mineral B. If mineral A and B both scratch each other, then their hardness is equal. A scale to measure hardness was devised by Austrian mineralogist Frederick (Friedrich) Mohs in 1822, and is the standard scale for measuring hardness. The scale consists of numbers one through ten; 1 being the softest and 10 being the hardest. Each number represents a different mineral - each harder than the previous. The 10 minerals are:

| 1. Talc | 2. Gypsum | 3. Calcite | 4. Fluorite | 5. Apatite |
| 6. Feldspar | 7. Quartz | 8. Topaz | 9. Corundum | 10. Diamond |

All conceivable minerals fit in this scale, since Talc is the softest known mineral and Diamond the hardest. To demonstrate how to use the scale, understand the following example: Suppose a mineral scratches Fluorite, but not Apatite, then it has a hardness between 4 and 5.

Several common household items have a fixed hardness, and can be used to test for hardness:

| Fingernail | $2\frac{1}{2}$ |
|---|---|
| Penny | 3 |
| Knife blade | $5\frac{1}{2}$ |
| Glass | $5\frac{1}{2}$ |
| Steel file | $6\frac{1}{2}$ |
| Streak plate (floor tile) | $6\frac{1}{2}$ |

Hardness is almost always rounded off to the nearest half number.

There are various hardness testing kits. One type consists of 10 metal rods, each one containing a fragment of one of the minerals in the Moh's scale. Another type consists of large, low cost specimens of the Moh's minerals, labeled and stored in a wooden compartment box. The Diamond is either absent or a chip attached to a small metal rod. (The Diamond is really unnecessary, since no minerals are between hardness 9 and 10.)

A mineral is struck with a metal rod or "testing mineral" to test its hardness. It is tested in the manner of the following example:

| Action | Conclusion |
|---|---|
| Mineral struck with rod or mineral number 4 (Fluorite) from the testing kit. Mineral gets scratched. | Mineral must be less than or equal to 4. |
| Mineral struck with rod or mineral number 2 (Gypsum). Mineral does not get scratched. | Mineral must be between 2 and 4. |

| Mineral struck with rod or mineral number 3 (Calcite). Mineral gets scratched. | Mineral must be between 2 and 3. |
|---|---|

Two minerals with equal hardness will scratch each other. This gives an advantage to the hardness testing kit that includes real minerals over rods. One can scratch the mineral from the kit instead of scratching a nice specimen. In addition, one can also get more exact results by seeing if both minerals scratch each other.

Minerals can be damaged and lose value if not scratched properly. If a mineral testing kit is composed of minerals (as opposed to rods), it is preferable for the testing kit mineral to be scratched over the specimen. If this cannot be done, than the specimen has to be scratched. This should be done in an area where a scratch will be less noticeable, since it will make a permanent mark.

Hardness can be easily detected without a "kit". All one needs to know is the hardness of certain items (including the ones mentioned above) and minerals in his collection. These can be used instead of purchasing a kit.

## How to Test using Hardness

Hardness testing is done by "scratching" one mineral with the other. To get the most accurate results, a sharp edge should be scratched against a smooth surface, on a small an area as possible. The scratch should not be conducted on a surface that is coated, chipped, or weathered, for it will give inaccurate results.

When a mineral is scratched, a permanent indentation is created. Powder of the softer mineral will come off, and it will cover the scratch area. This powder needs to be brushed away to see if the mineral really got scratched, or if the powder of the softer mineral that was swiped across the specimen being tested created a scratch-like marking. When minerals of similar hardness are scratched together, it is difficult to tell which mineral (if not both of them) is really getting scratched because of this.

Most minerals are *anisotropic* to a minor extent, meaning their hardness varies in different directions. Kyanite is famous for this habit. When scratched in one direction, it exhibits a hardness of 4 to 5. When struck from the perpendicular direction, it exhibits a hardness of 6 to 7. Kyanite is the only mineral exhibiting such strong anisotropism. In virtually all minerals, the anisotropism is so weak that it cannot be determined.

## Cleavage and Fracture

In most of the crystals, strength of bonding in all directions is not equal which will be liable to crack along crystallographic directions representing a fracture property by reflecting the fundamental structure which can be analytical. Regular flat faces which resemble growth faces like in calcite and mica are the results of perfect cleavage. Cleavage is the property of a mineral to break along planes where bond strength is low in the chemical bond. A cleavage is said to be imperfect or parting when it is less developed. Parallel sheets are produced when some minerals break along one dominant plane of cleavage while prism or blocks results from breakage along two or more planes

Table: Types of Cleavage

| Cleavage Type | Angles | Example |
|---|---|---|
| Basal | Cleaves in one direction | Micas |
| Cubic | Cleaves in three directions at 900 to one another | Rock salt (halite), galena, etc. |
| Rhombohedral | Cleaves in three directions parallel to the faces of the rhombohedron but not at 900 to one another | Calcite |
| Octahedral | Cleaves in four directions parallel to the faces of the octahedron | Diamond and fluorite |
| Dodecahedral | Cleaves in six directions parallel to the faces of the dodecahedron | Sphalerite |
| Prismatic | Cleaves in two directions parallel to the prismatic faces | Pyroxenes and amphiboles |
| Pyramidal | Cleaves in four directions parallel to the pyramidal faces | Scheelite |

## Fracture

Fracture describes the quality of the cleavage surface. Most minerals display either uneven or grainy fracture, conchoidal (curved, shell-like lines) fracture, or hackly (rough, jagged) fracture.

If the mineral contains no planes of weakness, it will break along random directions called fracture. Following kinds of fracture patterns are observed:

- Conchoidal fracture - breaks along smooth curved surfaces.
- Fibrous and splintery - similar to the way wood breaks.
- Hackly - jagged fractures with sharp edges.
- Uneven or Irregular - rough irregular surfaces.

## Parting

Parting is also a plane of weakness in the crystal structure, but it is along planes that are weakened by some applied force. It therefore may not be apparent in all specimens of the same mineral, but may appear if the mineral has been subjected to the right stress conditions.

## Types of Cleavage

Cleavage forms parallel to crystallographic planes:

- Basal or pinacoidal cleavage occurs when there is only one cleavage plane. Graphite has basal cleavage. Mica (like muscovite or biotite) also has basal cleavage; this is why mica can be peeled into thin sheets.

- Cubic cleavage occurs on when there are three cleavage planes intersecting at 90 degrees. Halite (or salt) has cubic cleavage, and therefore, when halite crystals are broken, it will form more cubes.

- Octahedral cleavage occurs when there are four cleavage planes in a crystal. Fluorite exhibits perfect octahedral cleavage. Octahedral cleavage is common for semiconductors. Diamond also has octahedral cleavage.

- Rhombohedral cleavage occurs when there are three cleavage planes intersecting at angles that are not 90 degrees. Calcite had rhombohedral cleavage.

- Prismatic cleavage occurs when there are two cleavage planes in a crystal. Spodumene exhibits prismatic cleavage.

- Dodecahedral cleavage occurs when there are six cleavage planes in a crystal. Sphalerite has dodecahedral cleavage.

## Parting

Crystal parting occurs when minerals break along planes of structural weakness due to external stress or along twin composition planes. Parting breaks are very similar in appearance to cleavage, but only occur due to stress. Examples include magnetite which shows octahedral parting, the rhombohedral parting of corundum and basal parting in pyroxenes.

## Uses

Cleavage is a physical property traditionally used in mineral identification, both in hand specimen and microscopic examination of rock and mineral studies. As an example, the angles between the prismatic cleavage planes for the pyroxenes (88–92°) and the amphiboles (56–124°) are diagnostic.

Crystal cleavage is of technical importance in the electronics industry and in the cutting of gemstones.

Precious stones are generally cleaved by impact, as in diamond cutting.

Synthetic single crystals of semiconductor materials are generally sold as thin wafers which are much easier to cleave. Simply pressing a silicon wafer against a soft surface and scratching its edge with a diamond scribe is usually enough to cause cleavage; however, when dicing a wafer to form chips, a procedure of scoring and breaking is often followed for greater control. Elemental semiconductors (Si, Ge, and diamond) are diamond cubic, a space group for which octahedral cleavage is observed. This means that some orientations of wafer allow near-perfect rectangles to be cleaved. Most other commercial semiconductors (GaAs, InSb, etc.) can be made in the related zinc blende structure, with similar cleavage planes.

## Difference between Cleavage and Fracture

In mineralogy, cleavage and fracture both describe a tendency of a mineral to break. In cleavage, a mineral may split apart along various smooth planes. These smooth planes are parallel to zones of weak bonding. On the other hand, fracture breaks a mineral along the curved surface with irregular shapes. The minerals that undergo fracture do not have planes of weakness, and therefore break in an irregular manner.

Cleavage describes a way in which a mineral breaks into flat surfaces. The number of flat surfaces usually range from one to four. This behavior is defined by the crystal structure of the mineral. Cleavage can be simply defined as the property of some minerals to break along specific planes in the crystal. The reason behind the planes of weakness is the weak bonding between the atoms that constitutes a mineral. Thus, cleavage can be associated with the atomic arrangement of a mineral.

Cleavage is often measured by three factors:

- Quality of cleavage
- Number of sides exhibiting cleavage
- Cleavage habit

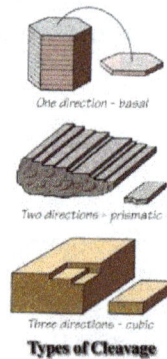

One direction - basal

Two directions - prismatic

Three directions - cubic

**Types of Cleavage**

It is also important to describe the angle between the cleavage planes. For example, if a mineral is in the shape of a cube then the angle between the cleavage planes should be 90°.

Fracture breaks a mineral along the curved surface with irregular shapes. The minerals that undergo fracture do not have planes of weakness, and therefore break in an irregular manner. Fracture exhibits a more random way of breaking than compared to cleavage. It occurs, as the bonds between all the atoms of the mineral are roughly equal. Thus, there are no layers of weakness like in the case of cleavage. Therefore, the mineral breaks in a random manner.

Fracture helps in the identification of the mineral, and therefore is considered to be one of the various properties including: streak, specific gravity, crystal structure, luster, color, hardness, flame test and others. Cleavage planes can be easily distinguished from a fracture as they are smooth and often have reflective surfaces.

Comparison between Cleavage and Fracture:

| | Cleavage | Fracture |
|---|---|---|
| Definition | In cleavage, a mineral may split apart along various smooth planes. These smooth planes are parallel to zones of weak bonding. | Fracture breaks a mineral along the curved surface with irregular shapes. |
| Types | There are five types of cleavage:<br>• Basal or pinacoidal – It occurs parallel to the base of a crystal<br>• Cubic Cleavage – It occurs parallel to the faces of a cube for a crystal with cubic symmetry.<br>• Octahedral Cleavage – It occurs on the 111 crystal planes and octahedral shapes are formed.<br>• Dodecahedral Cleavage - It occurs on the 110 crystal pleanes and dodecahedra shapes are formed.<br>• Rhombohedral cleavage – It occurs parallel to the 1011 faces of a rhombohedron.<br>• Prismatic cleavage - It is a cleavage that is parallel to a vertical prism 110. | There are five types of fracture:<br>• Conchoidal - A fracture resembling a semicircular shell, with a smooth, curved surface.<br>• Uneven - A fracture that leaves a rough or irregular surface.<br>• Hackly - A hackly fracture that resembles broken metals, with rough, jagged, points. True metals exhibit this fracture.<br>• Splintery - This type of fracture will form elongated splinters. All fibrous minerals fall into this category.<br>• Earthy or crumbly- This describes minerals that crumble when broken. |
| Examples | Halite, mica, and calcite | Quartz and obsidian |
| Reason | Inherent weaknesses within a mineral's atomic structures. | No planes of weakness in a mineral's atomic structure. |

## Crystal Form

Crystal form is the external expression of the internal ordered arrangement of atoms. During mineral formation, individual crystals develop well-formed crystal faces that are specific to that mineral. They reflect the internal symmetry of the crystal structure that makes the mineral unique. Crystal faces commonly seen on quartz are growth faces and represent the slowest growing directions in the structure. Quartz grows rapidly along its c-axis (three-fold or trigonal symmetry axis) direction and so never shows faces perpendicular to this direction. On the other hand, calcite rhomb faces and mica plates are cleavages and represent the weakest chemical bonds in the structure. The crystal faces for a particular mineral are characterized by a symmetrical relationship to one another that is manifest in the physical shape of the mineral's crystalline form. Crystal forms are commonly classified using six different crystal systems (Figure (a-f)), under which all minerals are grouped.

## Cubic or Isometric

Cubic or Isometric the simple cubic system has one lattice point on each corner of the cube with each lattice point shared equally between eight adjacent cubes. eg. Halite as rock salt.

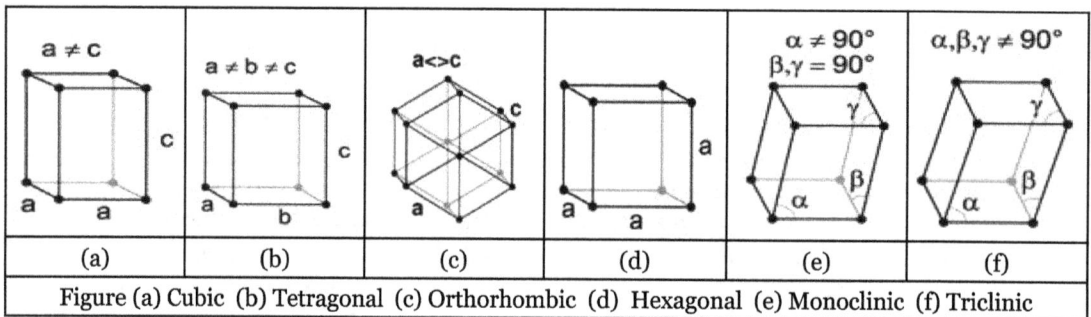

| (a) | (b) | (c) | (d) | (e) | (f) |
|-----|-----|-----|-----|-----|-----|

Figure (a) Cubic  (b) Tetragonal  (c) Orthorhombic  (d)  Hexagonal  (e) Monoclinic  (f) Triclinic

- Halite

  Halite, commonly known as rock salt, is a type of salt, the mineral (natural) form of sodium chloride ($NaCl$). Halite forms isometric crystals. The mineral is typically colorless or white, but may also be light blue, dark blue, purple, pink, red, orange, yellow or gray depending on the amount and type of impurities. It commonly occurs with other evaporite deposit minerals such as several of the sulfates, halides, and borates.

## Tetragonal

Tetragonal crystal lattices result from stretching a cubic lattice along one lattice vectors, making the cube a rectangular prism with a square base. e g. Zircon.

- Zircon

  Zircon is a mineral belonging to the group of nesosilicates. Its chemical name is zirconium silicate, and its corresponding chemical formula is $ZrSiO_4$. A common empirical formula showing some of the range of substitution in zircon is $(Zr_{1-y}, REE_y)(SiO_4)_{1-x}(OH)_{4x-y}$. Zircon forms in silicate melts with large proportions of high field strength incompatible elements. For example, hafnium is almost always present in quantities ranging from 1 to 4%. The crystal structure of zircon is tetragonal crystal system. The natural color of zircon varies between colorless, yellow-golden, red, brown, blue and green. Colorless specimens that show gem quality are a popular substitute for diamond and are also known as "Matura diamond".

The name derives from the Persian *zargun*, meaning "gold-hued". This word is corrupted into "jargoon", a term applied to light-colored zircons. The English word "zircon" is derived from *Zirkon*, which is the German adaptation of this word. Yellow, orange and red zircon is also known as "hyacinth", from the flower *hyacinthus*, whose name is of Ancient Greek origin.

## Orthorhombic

Orthorhombic lattices are made by stretching a cubic lattice along two lattice vectors by two factors, forming a rectangular prism with a rectangular base. All three bases intersect at 90 degree angles and the three lattice vectors are mutually orthogonal. e g. Aragonite.

- Aragonite

  Aragonite is a carbonate mineral, one of the two most common, naturally occurring, crystal forms of calcium carbonate, $CaCO_3$ (the other forms being the minerals calcite and vaterite). It is formed by biological and physical processes, including precipitation from marine and freshwater environments.

Aragonite's crystal lattice differs from that of calcite, resulting in a different crystal shape, an orthorhombic crystal system with acicular crystal. Repeated twinning results in pseudo-hexagonal forms. Aragonite may be columnar or fibrous, occasionally in branching stalactitic forms called *flos-ferri* ("flowers of iron") from their association with the ores at the Carinthian iron mines.

## Hexagonal

Hexagonal lattice has the same symmetry as a right prism with a hexagonal base. Graphite is an example of a hexagonal crystal.

- Graphite

  Graphite archaically referred to as plumbago, is a crystalline allotrope of carbon, a semimetal, a native element mineral, and a form of coal. Graphite is the most stable form of carbon under standard conditions. Therefore, it is used in thermochemistry as the standard state for defining the heat of formation of carbon compounds.

- Structure

  Graphite has a layered, planar structure. The individual layers are called graphene. In each layer, the carbon atoms are arranged in a honeycomb lattice with separation of 0.142 nm, and the distance between planes is 0.335 nm. Atoms in the plane are bonded covalently, with only three of the four potential bonding sites satisfied. The fourth electron is free to migrate in the plane, making graphite electrically conductive. However, it does not conduct in a direction at right angles to the plane. Bonding between layers is via weak van der Waals bonds, which allows layers of graphite to be easily separated, or to slide past each other.

The two known forms of graphite, *alpha* (hexagonal) and *beta* (rhombohedral), have very similar physical properties, except for that the graphene layers stack slightly differently. The alpha graphite may be either flat or buckled. The alpha form can be converted to the beta form through mechanical treatment and the beta form reverts to the alpha form when it is heated above 1300 °C.

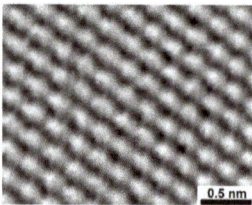

Scanning tunneling microscope image of graphite surface atoms

Graphite's unit cell

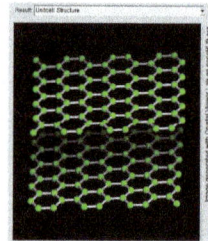

Ball-and-stick model of graphite (two graphene layers)

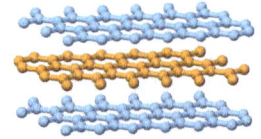

Side view of layer stacking

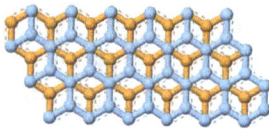

Plane view of layer stacking

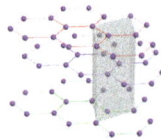

Animated view of the unit cell in three layers of graphene (note that this is a slightly different unit cell from the one to the left)

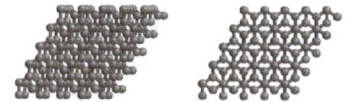

Rotating graphite stereogram

## Monoclinic

Monoclinic lattice is described by vectors of unequal length that form a rectangular prism with a parallelogram as base. Two pairs of vectors are perpendicular, while the third pair makes an angle other than 90 degrees. e g. Orthoclase

- Orthoclase

Orthoclase, or orthoclase feldspar (endmember formula $KAlSi_{3}O_{8}$), is an important tectosilicate mineral which forms igneous rock. The name is from the Ancient Greek for "straight fracture," because its two cleavage planes are at right angles to each other. It is a type of potassium feldspar, also known as K-feldspar. The gem known as moonstone is largely composed of orthoclase.

## Triclinic

Triclinic crystal is described by vectors of unequal length. All three vectors are not mutually orthogonal. e g. Copper sulphate.

## Copper(I) Sulfate

Copper(I) sulfate, also known as cuprous sulfate and dicopper sulfate, is the chemical compound with the chemical formula $Cu_2SO_4$ and a molar mass of 223.15 g mol$^{-1}$. It is an unstable compound as oxo-acids are generally unstable and is more commonly found in the $CuSO_4$ state. It is light green in color at room temperature and is water-soluble. Due to the low-stability of the compound there are currently not many applications to date.

## Structure

$Cu_2SO_4$ crystallizes in the orthorhombic space group Fddd with cubic closest pack cell unit properties a = 474.8, b = 1396, and c = 1086 pm, Z = 8, Dx = 4.12 g cm$^{-3}$. The structure is formed by the tendency of Cu(I) to form two collinear sp bonds to oxygen atoms which result in short O–Cu–O groups with the O–Cu bond length being approximately 196vpm. Due to the bonding properties of the Cu(I) metal the structure is built of layers of composition $Cu_2SO_4$. Due to the loss of only one electron and the bonding properties of copper(I) the structure is built up from four oxygen atoms of each sulfate group bonding to four other sulfate groups of the same layer via symmetrical O–Cu–O bridging. The layer structure of $Cu_2SO_4$ is built as a consequence of the formation of unsymmetrical O–H–O hydrogen bridges. Due to its unstable nature it is difficult to obtain an accurate 3D structure, but was first successfully done in 1987 by Berthold, Born, and Wartchow.

## Copper(II) Sulfate

Copper(II) sulfate, also known as cupric sulfate, or copper sulphate, is the inorganic compound with the chemical formula $CuSO_4(H_2O)_x$, where x can range from 0 to 5. The pentahydrate (x = 5) is the most common form. Older names for this compound include blue vitriol, bluestone, vitriol of copper, and Roman vitriol.

The pentahydrate ($CuSO_4 \cdot 5H_2O$), the most commonly encountered salt, is bright blue. It exothermically dissolves in water to give the aquo complex $[Cu(H_2O)_6]^{2+}$, which has octahedral molecular geometry. The structure of the solid pentahydrate reveals a polymeric structure wherein copper is again octahedral but bound to four water ligands. The $Cu(II)(H_2O)_4$ centers are interconnected by sulfate anions to form chains. Anhydrous copper sulfate is a white powder.

## Tenacity

Tenacity is the tendency of a mineral to deform plastically under stress. Minerals may be brittle and can be fractured under stress such as silicates and oxides. Minerals can be sectile as they can be cut with a knife or they can be ductile by deforming readily under stress as does gold. The following chart below gives the list of terms used to describe tenacity and a description of each (Table)

| Table: Various terms related to Tenacity | | |
|---|---|---|
| **Term** | **Description** | **Example** |
| Sectile | Mineral can be cut with a knife resulting shice breaks up under a hammer. | Graphite, steatite, gypsum |
| Maalleable | A slice cut from mineral can be hammered out into thin flat sheets | Native gold, silver, copper |
| Flexible | When applied pressure a mineral can bent but does not return to its original position, when the pressure is removed | |
| Elastic | The mineral or thin plates of laminac from it can bent and returns to its original position after the pressure is removed | Micas |
| Brittle | The mineral crumble or shatters easily | Iron, pyrite, apatite, fluorite |
| Ductile | The mineral can be drawn into wires | Copper |

Brittle-If a mineral is hammered and the result is a powder or small crumbs, it is considered brittle. Brittle minerals leave a fine powder if scratched, which is the way to test a mineral to see if it is brittle. The majority of all minerals are brittle. An example is Quartz. (Minerals that are not brittle may be referred to as Nonbrittle minerals.)

Sectile - Sectile minerals can be separated with a knife, much like wax but usually not as soft.

An example is Gypsum.

Malleable - If a mineral can be flattened by pounding with a hammer, it is malleable. All true metals are malleable.

An example is Silver.

Ductile - A mineral that can be stretched into a wire is ductile. All true metals are ductile.

An example is Gold.

Flexible but inelastic - Any mineral that can be bent, but remains in the new position after it is bent is flexible but inelastic. If the term flexible is singularly used, it implies flexible but inelastic.

An example is Copper.

Flexible and elastic - When flexible and elastic minerals are bent, they spring back to their original position. All fibrous minerals, and some acicular minerals belong in this category. An example is Chrysotile Serpentine.

## How to use Tenacity as an Identification Mark

Each form of tenacity is tested on its own. A mineral of questionable status should not be tested for the different forms of tenacity, except for brittleness; such tests can damage a specimen. Minerals are not tested for malleability or ductility, since the only way to conduct these tests is by pounding and stretching the mineral.

## How to Test using Tenacity

When testing for brittleness, one should scratch the specimen with a harder mineral or material, and see if the mineral leaves a fine powder. If fine powder is scratched off the mineral, it is brittle. If just a furrow is left without any fine powder, the mineral is nonbrittle. Brittleness and hardness are usually tested simultaneously, since the same procedure can be conducted for both of them. When testing for brittleness, do not scratch a noticeable surface area, as this may decrease a specimens value. Additionally, the surface being tested by scratching should not be dirty, coated, or tarnished.

If a mineral is suspected as being sectile, insert a knife into the mineral and see if it goes through. If the knife penetrates, the mineral is sectile. Sectility can also be tested with a long fingernail, by slowly pushing it into the specimen to see if it enters. If the fingernail goes through, the mineral is sectile.

To check for flexibility, slight pressure should be placed on the mineral to see if it bends. If it moves to the new position after stress is released, it is inelastic. If it goes back to its original position after stress is released, it is elastic. Even some flexible minerals will break if there is too much stress is applied. Therefore, they should be bent carefully and slowly, and only a minimum amount. Only few minerals are flexible, so thin minerals may be damaged even when slight flexing pressure is applied.

## Luster or Transparency

*Luster* refers to the general appearance of a mineral surface in reflected light. Mineral luster can be metallic or non-metallic. The mineral is metallic when it is opaque and reflects like metals. Minerals with metallic luster are usually opaque and have a colored streak .Non-metallic luster can be :

    1. vitreous eg. Quartz, calcite

    2. resinous eg. Sulfur

    3. pearly eg. Barite

    4. greasy eg. Calcite

    5. silky eg. Asbestos

    6. Adamantine eg. Feldspar

Alternatively, it is non-metallic when mineral does not reflect light. When the mineral looks like paraffin or wax then it is said to be waxy. Examples include jade and chalcedony . The mineral is vitreous when it looks like broken glass and it is pearly when appears iridescent for example muscovite and stilbite. The mineral looks fibrous like silk then it is termed as silky. Examples of silky minerals include asbestos, ulexite and the satin spar variety of gypsum. The mineral is greasy when looks like oil on water. Resinous minerals looks like hardened tree sap (resin), example is amber. Admantine looks brilliant, like diamond. Greasy minerals resemble fat or grease. A greasy lustre often occurs in minerals containing a great abundance of microscopic inclusions for example opal and cordierite.

## Adamantine lustre

Cut diamonds

Adamantine minerals possess a superlative lustre, which is most notably seen in diamond. Such minerals are transparent or translucent, and have a high refractive index (of 1.9 or more). Minerals with a true adamantine lustre are uncommon, with examples being cerussite and cubic zirconia.

Minerals with a lesser (but still relatively high) degree of lustre are referred to as subadamantine, with some examples being garnet and corundum.

## Dull Lustre

Kaolinite

Dull (or earthy) minerals exhibit little to no lustre, due to coarse granulations which scatter light in all directions, approximating a Lambertian reflector. An example is kaolinite. A distinction is sometimes drawn between dull minerals and earthy minerals, with the latter being coarser, and having even less lustre.

## Greasy Lustre

Moss opal

Greasy minerals resemble fat or grease. A greasy lustre often occurs in minerals containing a great abundance of microscopic inclusions, with examples including opal and cordierite. Many minerals with a greasy lustre also feel greasy to the touch.

## Metallic Lustre

Pyrite

Metallic (or splendent) minerals have the lustre of polished metal, and with ideal surfaces will work as a reflective surface. Examples include galena, pyrite and magnetite.

## Pearly Lustre

Muscovite

Pearly minerals consist of thin transparent co-planar sheets. Light reflecting from these layers give them a lustre reminiscent of pearls. Such minerals possess perfect cleavage, with examples including muscovite and stilbite.

## Resinous Lustre

Amber

Resinous minerals have the appearance of resin, chewing gum or (smooth-surfaced) plastic. A principal example is amber, which is a form of fossilized resin.

## Silky Lustre

Satin spar variety of gypsum

Silky minerals have a parallel arrangement of extremely fine fibres, giving them a lustre reminiscent of silk. Examples include asbestos, ulexite and the satin spar variety of gypsum. A fibrous lustre is similar, but has a coarser texture.

## Submetallic Lustre

Sphalerite

Submetallic minerals have similar lustre to metal, but are duller and less reflective. A submetallic lustre often occurs in near-opaque minerals with very high refractive indices, such as sphalerite, cinnabar and cuprite.

## Vitreous lustre

Quartz

Vitreous minerals have the lustre of glass. (The term is derived from the Latin for glass, *vitrum*.) This type of lustre is one of the most commonly seen, and occurs in transparent or translucent minerals with relatively low refractive indices. Common examples include calcite, quartz, topaz, beryl, tourmaline and fluorite, among others.

## Waxy Lustre

Jade

Waxy minerals have a lustre resembling wax. Examples include jade and chalcedony.

## Optical Phenomena

## Asterism

Sapphire cabochon

Asterism is the display of a star-shaped luminous area. It is seen in some sapphires and rubies, where it is caused by impurities of rutile. It can also occur in garnet, diopside and spinel.

## Aventurescence

Aventurine

Aventurescence (or aventurization) is a reflectance effect like that of glitter. It arises from minute, preferentially oriented mineral platelets within the material. These platelets are so numerous that they also influence the material's body colour. In aventurine quartz, chrome-bearing fuchsite makes for a green stone and various iron oxides make for a red stone.

## Chatoyancy

Tiger's eye

Chatoyant minerals display luminous bands, which appear to move as the specimen is rotated. Such minerals are composed of parallel fibers (or contain fibrous voids or inclusions), which reflect light into a direction perpendicular to their orientation, thus forming narrow bands of light. The most famous examples are tiger's eye and cymophane, but the effect may also occur in other minerals such as aquamarine, moonstone and tourmaline.

## Color Change

Alexandrite

Color change is most commonly found in alexandrite, a variety of chrysoberyl gemstones. Other gems also occur in color-change varieties, including (but not limited to) sapphire, garnet, spinel. Alexandrite displays a color change dependent upon light, along with strong pleochroism. The gem results from small-scale replacement of aluminium by chromium oxide, which is responsible for alexandrite's characteristic green to red color change. Alexandrite from the Ural Mountains in Russia is green by daylight and red by incandescent light. Other varieties of alexandrite may be yellowish or pink in daylight and a columbine or raspberry red by incandescent light. The optimum or "ideal" color change would be fine emerald green to fine purplish red, but this is exceedingly rare.

## Schiller

Labradorite

Schiller, from German for "color play", is the metallic iridescence originating from below the surface of a stone that occurs when light is reflected between layers of minerals. It is seen in moonstone and labradorite and is very similar to adularescence and aventurescence.

## Specific Gravity

Specific gravity is a definite physical property calculated in g/cm3. It can be calculated by the measurement of volume with displacement of water in a graduated cylinder and the mass. Specific gravity is a dimensionless quantity. Different minerals have different Specific gravity e.g. Iron (8), lead (13) and gold (19). The range of specific gravity is from 2.6-3.5 for silicates and from 5-6 for sulfides. Mineral with various specific gravity are summarized in Table. Specific gravity (SG) is calculated by the determination of the weight in air (Wa) and the weight in water (Ww) and specific gravity can be calculated using following formula:

SG = Wa / (Wa-Ww).

Table: Mineral with various Specific gravity

| Mineral | Composition | Specific Gravity |
|---------|-------------|------------------|
| Graphite | C | 2.23 |
| Quartz | $SiO_2$ | 2.65 |

| Feldspars | (K,Na)AlSi3O8 | 2.6 - 2.75 |
|---|---|---|
| Fluorite | CaF2 | 3.18 |
| Topaz | Al2SiO4(F,OH)2 | 3.53 |
| Corundum | Al2O3 | 4.02 |
| Barite | BaSO4 | 4.45 |
| Pyrite | FeS2 | 5.02 |
| Galena | PbS | 7.5 |
| Cinnabar | HgS | 8.1 |
| Copper | Cu | 8.9 |
| Silver | Ag | 10.5 |

## Details

Being a ratio of densities, specific gravity is a dimensionless quantity. The reason for the specific gravity being dimensionless is to provide a global consistency between the U.S. and Metric Systems, since various units for density may be used such as pounds per cubic feet or grams per cubic centimeter, etc. Specific gravity varies with temperature and pressure; reference and sample must be compared at the same temperature and pressure or be corrected to a standard reference temperature and pressure. Substances with a specific gravity of 1 are neutrally buoyant in water. Those with SG greater than 1 are denser than water and will, disregarding surface tension effects, sink in it. Those with an SG less than 1 are less dense than water and will float on it. In scientific work, the relationship of mass to volume is usually expressed directly in terms of the density (mass per unit volume) of the substance under study. It is in industry where specific gravity finds wide application, often for historical reasons.

True specific gravity can be expressed mathematically as:

$$SG_{true} = \frac{\rho_{sample}}{\rho_{H_2O}}$$

where $\rho_{sample}$ is the density of the sample and $\rho_{H2O}$ is the density of water.

The apparent specific gravity is simply the ratio of the weights of equal volumes of sample and water in air:

$$SG_{apparent} = \frac{W_{A,sample}}{W_{A,H_2O}}$$

where $W_{A,sample}$ represents the weight of the sample measured in air and $W_{A,H2O}$ the weight of water measured in air.

It can be shown that true specific gravity can be computed from different properties:

$$SG_{true} = \frac{\rho_{sample}}{\rho_{H_2O}} = \frac{\dfrac{m_{sample}}{V}}{\dfrac{m_{H_2O}}{V}} = \frac{m_{sample}}{m_{H_2O}}\frac{g}{g} = \frac{W_{V,sample}}{W_{V,H_2O}}$$

where $g$ is the local acceleration due to gravity, $V$ is the volume of the sample and of water (the same for both), $\rho_{sample}$ is the density of the sample, $\rho_{H2O}$ is the density of water and $W_V$ represents a weight obtained in vacuum.

The density of water varies with temperature and pressure as does the density of the sample. So it is necessary to specify the temperatures and pressures at which the densities or weights were determined. It is nearly always the case that measurements are made at 1 nominal atmosphere (101.325 kPa ± variations from changing weather patterns). But as specific gravity usually refers to highly incompressible aqueous solutions or other incompressible substances (such as petroleum products), variations in density caused by pressure are usually neglected at least where apparent specific gravity is being measured. For true (*in vacuo*) specific gravity calculations, air pressure must be considered. Temperatures are specified by the notation $(T_s/T_r)$, with $T_s$ representing the temperature at which the sample's density was determined and $T_r$ the temperature at which the reference (water) density is specified. For example, SG (20 °C/4 °C) would be understood to mean that the density of the sample was determined at 20 °C and of the water at 4 °C. Taking into account different sample and reference temperatures, we note that, while $SG_{H2O}$ = 1.000000 (20 °C/20 °C), it is also the case that $SG_{H2O}$ = $\frac{0.998203}{0.999840}$ = 0.998363 (20 °C/4 °C). Here, temperature is being specified using the current ITS-90 scale and the densities used here and in the rest of this article are based on that scale. On the previous IPTS-68 scale, the densities at 20 °C and 4 °C are 0.9982071 and 0.9999720 respectively, resulting in an SG (20 °C/4 °C) value for water of 0.9982343.

As the principal use of specific gravity measurements in industry is determination of the concentrations of substances in aqueous solutions and as these are found in tables of SG versus concentration, it is extremely important that the analyst enter the table with the correct form of specific gravity. For example, in the brewing industry, the Plato table lists sucrose concentration by weight against true SG, and was originally (20 °C/4 °C) i.e. based on measurements of the density of sucrose solutions made at laboratory temperature (20 °C) but referenced to the density of water at 4 °C which is very close to the temperature at which water has its maximum density $\rho_{H2O}$ equal to 999.972 kg/m³ in SI units ({{val|0.999972|u=g/cm³ in cgs units or 62.43 lb/cu ft in United States customary units). The ASBC table in use today in North America, while it is derived from the original Plato table is for apparent specific gravity measurements at (20 °C/20 °C) on the IPTS-68 scale where the density of water is 0.9982071 g/cm³. In the sugar, soft drink, honey, fruit juice and related industries sucrose concentration by weight is taken from a table prepared by A. Brix which uses SG (17.5 °C/17.5 °C). As a final example, the British SG units are based on reference and sample temperatures of 60 °F and are thus (15.56 °C/15.56 °C).

Given the specific gravity of a substance, its actual density can be calculated by rearranging the above formula:

$$\rho_{substance} = SG \times \rho_{H_2O}.$$

Occasionally a reference substance other than water is specified (for example, air), in which case specific gravity means density relative to that reference.

## Measurement: Apparent and True Specific Gravity

### Pycnometer

Specific gravity can be measured in a number of value ways. The following illustration involving the use of the pycnometer is instructive. A pycnometer is simply a bottle which can be precisely filled to a specific, but not necessarily accurately known volume, $V$. Placed upon a balance of some sort it will exert a force.

$$F_b = g\left( m_b - \rho_a \frac{m_b}{\rho_b} \right)$$

where $m_b$ is the mass of the bottle and $g$ the gravitational acceleration at the location at which the measurements are being made. $\rho_a$ is the density of the air at ambient pressure and $\rho_b$ is the density of the material of which the bottle is made (usually glass) so that the second term is the mass of air displaced by the glass of the bottle whose weight, by Archimedes Principle must be subtracted. The bottle is filled with air, but as that air displaces an equal amount of air the weight of that air is canceled by the weight of the air displaced. Now we fill the bottle with the reference fluid, for example pure water. The force exerted on the pan of the balance becomes:

$$F_w = g\left( m_b - \rho_a \frac{m_b}{\rho_b} + V\rho_w - V\rho_a \right).$$

If we subtract the force measured on the empty bottle from this (or tare the balance before making the water measurement) we obtain.

$$F_{w,n} = gV(\rho_w - \rho_a)$$

where the subscript n indicates that this force is net of the force of the empty bottle. The bottle is now emptied, thoroughly dried and refilled with the sample. The force, net of the empty bottle, is now:

$$F_{s,n} = gV(\rho_s - \rho_a)$$

where $\rho_s$ is the density of the sample. The ratio of the sample and water forces is:

$$SG = \frac{gV(\rho_s - \rho_a)}{gV(\rho_w - \rho_a)} = \frac{\rho_s - \rho_a}{\rho_w - \rho_a}$$

This is called the Apparent Specific Gravity, denoted by subscript A, because it is what we would obtain if we took the ratio of net weighings in air from an analytical balance or used a hydrometer (the stem displaces air). Note that the result does not depend on the calibration of the balance. The only requirement on it is that it read linearly with force. Nor does $SG_A$ depend on the actual volume of the pycnometer.

Further manipulation and finally substitution of $SG_V$, the true specific gravity (the subscript V is used because this is often referred to as the specific gravity *in vacuo*), for $\rho_{s/\rho w}$ gives the relationship between apparent and true specific gravity.

$$SG_A = \frac{\dfrac{\rho_s}{\rho_w} - \dfrac{\rho_a}{\rho_w}}{1 - \dfrac{\rho_a}{\rho_w}} = \frac{SG_V - \dfrac{\rho_a}{\rho_w}}{1 - \dfrac{\rho_a}{\rho_w}}$$

In the usual case we will have measured weights and want the true specific gravity. This is found from

$$SG_V = SG_A - \frac{\rho_a}{\rho_w}(SG_A - 1).$$

Since the density of dry air at 101.325 kPa at 20 °C is 0.001205 g/cm³ and that of water is 0.998203 g/cm³ the difference between true and apparent specific gravities for a substance with specific gravity (20 °C/20 °C) of about 1.100 would be 0.000120. Where the specific gravity of the sample is close to that of water (for example dilute ethanol solutions) the correction is even smaller.

## Digital Density Meters

- Hydrostatic pressure-based instruments

  This technology relies upon Pascal's Principle which states that the pressure difference between two points within a vertical column of fluid is dependent upon the vertical distance between the two points, the density of the fluid and the gravitational force. This technology is often used for tank gauging applications as a convenient means of liquid level and density measure.

- Vibrating element transducers

This type of instrument requires a vibrating element to be placed in contact with the fluid of interest. The resonant frequency of the element is measured and is related to the density of the fluid by a characterization that is dependent upon the design of the element. In modern laboratories precise measurements of specific gravity are made using oscillating U-tube meters. These are capable of measurement to 5 to 6 places beyond the decimal point and are used in the brewing, distilling, pharmaceutical, petroleum and other industries. The instruments measure the actual mass of fluid contained in a fixed volume at temperatures between 0 and 80 °C but as they are microprocessor based can calculate apparent or true specific gravity and contain tables relating these to the strengths of common acids, sugar solutions, etc. The vibrating fork immersion probe is another good example of this technology. This technology also includes many coriolis-type mass flow meters which are widely used in chemical and petroleum industry for high accuracy mass flow measurement and can be configured to also output density information based on the resonant frequency of the vibrating flow tubes.

- Ultrasonic transducer

Ultrasonic waves are passed from a source, through the fluid of interest, and into a detector which measures the acoustic spectroscopy of the waves. Fluid properties such as density and viscosity can be inferred from the spectrum.

- Radiation-based gauge

Radiation is passed from a source, through the fluid of interest, and into a scintillation detector, or counter. As the fluid density increases, the detected radiation "counts" will decrease. The source is typically the radioactive isotope cesium-137, with a half-life of about 30 years. A key advantage for this technology is that the instrument is not required to be in contact with the fluid – typically the source and detector are mounted on the outside of tanks or piping.

- Buoyant force transducer

The buoyancy force produced by a float in a homogeneous liquid is equal to the weight of the liquid that is displaced by the float. Since buoyancy force is linear with respect to the density of the liquid within which the float is submerged, the measure of the buoyancy force yields a measure of the density of the liquid. One commercially available unit claims the instrument is capable of measuring specific gravity with an accuracy of ±0.005 SG units. The submersible probe head contains a mathematically characterized spring-float system. When the head is immersed vertically in the liquid, the float moves vertically and the position of the float controls the position of a permanent magnet whose displacement is sensed by a concentric array of Hall-effect linear displacement sensors. The output signals of the sensors are mixed in a dedicated electronics module that provides an output voltage whose magnitude is a direct linear measure of the quantity to be measured.

- Inline continuous measurement

Slurry is weighed as it travels through the metered section of pipe using a patented, high resolution load cell. This section of pipe is of optimal length such that a truly representative mass of the slurry may be determined. This representative mass is then interrogated by the load cell 110 times per second to ensure accurate and repeatable measurement of the slurry.

## Geological Processes of Minerals Formation

### Magmatic Deposits

Magmatic deposits result from simple crystallization and concentration by differentiation of intrusive igneous masses. They have high melting points so that they can co-exist and get crystallized from silicate melts at temperatures above 800° C. Granites, granodiorites, and rhyolites, along with rich minerals like quartz, muscovite and alkali feldspars are Felsic igneous rocks. These minerals are usually light in color and the color is not always diagnostic. Pegmatite (PEG) is a third mineral environment showing concluding stages of fractionation in igneous rocks. It is very coarse grained and similar to silicic igneous rock in composition with high silica. There are certain elements which readily do not substitute in abundant minerals and are termed as incompatible elements. In pegmatites, they made their own minerals by accumulation. Minerals containing the incompatible elements, Li, Be, B, P, Rb, Sr, Y, Nb, rare earths, Cs, and Ta are typical and characteristic of pegmatites (Deer *et al.*, 1962, 1974, 1980). Rocks composed mostly of pyroxene, calcium-rich plagioclase, and minor amounts of olivine make up the mafic family of igneous rocks. The mafic magmas are somewhat more viscous than the ultramafic magmas, but they are still fairly fluid. The low silica and gas contents make ultramafic very fluid; i.e., they have a low viscosity, or resistance to flow. Ultramafic rocks are given names depending on whether they are intrusive or extrusive. Peridotite is the name given to intrusive ultramafic rocks, whereas komatiite is the name given to extrusive ultramafic rocks. Peridotite and komatiite are compositionally identical. Their textures, however, are different reflecting their mode of formation. Peridotite appears to be the dominant rock type of the upper mantle.

### Sedimentary Mineral Deposits

The formation of sedimentary rocks is accompanied by three processes. The first process is the weathering which produces the materials that a sedimentary rock is composed of by mechanical (freezing, thawing) and chemical (dissolution of minerals, formation of new minerals [clays]) interaction between atmosphere, hydrosphere and earth surface rocks. The second process is the transport which moves these materials to their final destination.

Rivers are the main transporting agent of material to the oceans (glaciers are at times important). During transport the sediment particles will be sorted according to size and density (gold placers) and will be rounded by abrasion. Material that has been dissolved during weathering will be carried away in solution. Winds may also play a role (Sahara -- east/central Atlantic). The sorting during transport is important because it is the reason that we have distinct clastic rock types (conglomerates, sandstones, shales). The third process is the deposition of sediment which occurs at a site with a specific combination of physical, chemical and biological conditions i.e. the sedimentary environment.

## Metamorphic Rock Mineral

Metamorphic processes occur to make adjustments between the chemical potential of any system and the changes in temperature and pressure. Metamorphism involves both chemical and mechanical changes but in varying proportions. Metamorphic minerals and rocks provide many valuable resources, marble and slate the two most widely used. Metamorphic rocks are formed when the precursor materials (igneous, sediment, etc.) are buried deeply and are consequently brought into an environment of high pressure and temperature. Therefore, they are most commonly encountered in the core zones of mountain belts (uplifted root zone), in old continental shields, and as the basement rock below the sediment veneer of stable continental platforms. Metamorphic rocks and associated igneous intrusions (from rock buried so deep that it melted) make up about 85% of the continental crust. Metamorphic rocks may contain relic structures, such as stratification, bedding, and even such features as sedimentary structures or volcanic textures.

## Hydrothermal Mineral Deposits

Hydrothermal mineral deposit is the fourth major mineral environments. The elements it contains and the minerals formed by this process are very different from regional and contact metamorphic rocks therefore; it becomes inevitable to consider them as isolated group. With increase in depth, there is a scarcity of water so minerals in the form of magma approach to the earth's surface through crust. Woods Hole Oceanographic Institute (USA) explored the floor of ocean with the use of submersible craft and discovered the hot waters plumes which were coming out along the oceanic ridge. The plumes of hot water contain manganese, copper, iron and zinc which are dissolved sulfide metals and the mineral deposits produced by them were termed massive sulfides. When viewed in the lights of submersible craft the water coming from tall hydrothermal vent resembled the smoke coming from chimney due to precipitated minerals and the vents were named as "black smokers". These may be sub-classified as high temperature hydrothermal (HTH), low temperature hydrothermal (LTH), and oxidized hydrothermal (OXH). Sulphide minerals generally consist of minerals which are present centrally and on the right hand side in the periodic table (e.g. Cu, Ag, As, Sb Sn, Pb, Zn, Hg, Cd) and are generally known as chalcophile elements. Sulfides are mostly hydrothermal but they may occur in metamorphic and igneous rocks. Silver, tungstate minerals, the tellurides chalcopyrite, gold, bornite, and molybdenite are all high temperature hydrothermal minerals and the hydrothermal minerals with low temperature include pyrite, barite, cinnabar, gold, and cassiterite. Sulfide minerals will weather by oxidation to form oxides of the sulfates and carbonates of the chalcophile metals as they are not stable in atmospheric oxygen. These types of mineral are the feature of oxidized hydrothermal deposits and are known as gossans. They show characterized mark i.e. red- yellow stains of iron oxide on the surface of rock. They generally separate mineralized zones at depth and are very common in Colorado.

## References

- Rafferty, John P. (2012). Geological sciences (1st ed.). New York: Britannica Educational Pub. in association with Rosen Educational Services. pp. 14–15. ISBN 9781615304950

- Scurfield, Gordon (1979). "Wood Petrifaction: an aspect of biomineralogy". Australian Journal of Botany. 27 (4): 377–390. doi:10.1071/bt9790377

- "Law of the constancy of interfacial angles". Online dictionary of crystallography. International Union of Crystallography. 24 August 2014. Retrieved 22 September 2015

- Scurfield, Gordon (1979). "Wood Petrifaction: an aspect of biomineralogy". Australian Journal of Botany. 27 (4): 377–390. doi:10.1071/bt9790377

- Holden, Martin (1991). The Encyclopedia of Gemstones and Minerals. New York: Facts on File. p. 251. ISBN 1-56799-949-2

- Ralph Nielsen "Zirconium and Zirconium Compounds" in Ullmann's Encyclopedia of Industrial Chemistry, 2005, Wiley-VCH, Weinheim. doi:10.1002/14356007.a28_543

- Lavrakas, Vasilis (1957). "Textbook errors: Guest column. XII: The lubricating properties of graphite". Journal of Chemical Education. 34 (5): 240. Bibcode:1957JChEd..34..240L. doi:10.1021/ed034p240

- Borenstein, Seth (19 October 2015). "Hints of life on what was thought to be desolate early Earth". Excite. Yonkers, NY: Mindspark Interactive Network. Associated Press. Retrieved 2015-10-20

- Deprez, N.; McLachlan, D. S. (1988). "The analysis of the electrical conductivity of graphite conductivity of graphite powders during compaction". Journal of Physics D: Applied Physics. Institute of Physics. 21 (1): 101–107. Bibcode:1988JPhD...21..101D. doi:10.1088/0022-3727/21/1/015

# Permissions

All chapters in this book are published with permission under the Creative Commons Attribution Share Alike License or equivalent. Every chapter published in this book has been scrutinized by our experts. Their significance has been extensively debated. The topics covered herein carry significant information for a comprehensive understanding. They may even be implemented as practical applications or may be referred to as a beginning point for further studies.

We would like to thank the editorial team for lending their expertise to make the book truly unique. They have played a crucial role in the development of this book. Without their invaluable contributions this book wouldn't have been possible. They have made vital efforts to compile up to date information on the varied aspects of this subject to make this book a valuable addition to the collection of many professionals and students.

This book was conceptualized with the vision of imparting up-to-date and integrated information in this field. To ensure the same, a matchless editorial board was set up. Every individual on the board went through rigorous rounds of assessment to prove their worth. After which they invested a large part of their time researching and compiling the most relevant data for our readers.

The editorial board has been involved in producing this book since its inception. They have spent rigorous hours researching and exploring the diverse topics which have resulted in the successful publishing of this book. They have passed on their knowledge of decades through this book. To expedite this challenging task, the publisher supported the team at every step. A small team of assistant editors was also appointed to further simplify the editing procedure and attain best results for the readers.

Apart from the editorial board, the designing team has also invested a significant amount of their time in understanding the subject and creating the most relevant covers. They scrutinized every image to scout for the most suitable representation of the subject and create an appropriate cover for the book.

The publishing team has been an ardent support to the editorial, designing and production team. Their endless efforts to recruit the best for this project, has resulted in the accomplishment of this book. They are a veteran in the field of academics and their pool of knowledge is as vast as their experience in printing. Their expertise and guidance has proved useful at every step. Their uncompromising quality standards have made this book an exceptional effort. Their encouragement from time to time has been an inspiration for everyone.

The publisher and the editorial board hope that this book will prove to be a valuable piece of knowledge for students, practitioners and scholars across the globe.

# Index